British Forestry in the 20th Century
Policy and Achievements

E.G. RICHARDS

Brill
Leiden–Boston
2003

This book is printed on acid-free paper

ISBN 90-6764-360-2

First published in 2003

Printed in The Netherlands by Ridderprint bv, Ridderkerk.

Contents

Acknowledgements

I should like to thank the members of the Forestry Commission staff who have answered a wide variety of questions which have arisen during the preparation of this book.

Thanks are due to the Commission for providing the photographs reproduced here. Line drawings and charts are the work of Peter Reddy.

Rosalynne Carter has typed the text from my original handwritten manuscript.

Professor Hugh Miller and Tim J. Peck kindly read the book in draft form and made valuable sugestions as to how it might be modified and I am grateful to them for their advice and help. Tim Peck also supplied the Foreword.

My thanks also go to John W.L. Zehetmayr, who read and corrected the proofs—a very substantial task from which the book has greatly benefited.

I thank Jan van der Linde, Senior Acquisition Editor, VSP International Science Publishers, whose experience has been of key importance in preparing the text for publication and in the printing of the book.

Notes on the Annual Reports of the Forestry Commissioners

The most commonly quoted source are the Annual Reports of the Forestry Commissioners to Parliament. The reports are referred to in the text and in the bibliography at the end of each chapter by the abbreviation CAR and either the number of the Report or its date or both.

The First Annual Report of the Forestry Commissioners was for the year ending 30th September 1920. Except for the war years 1939–1944 the reports have been published by HMSO.

These report on the details of the Commission's activities for the year under review. They include what is commonly called its Forest Authority role and its role as the Forest Enterprise managing the Commission's forest estate and other land; they also cover the various Forestry Acts and other legislation impinging on the Commission's activities.

From the first (for the year ended 30th September 1920) to the forty-seventh, they were titled Forestry Commission, First Annual Report of the Forestry Commissioners; Second Annual Report of the Forestry Commissioners; and so on.

From the forty-eighth, for the year ended 30th September 1967, the title changed to Forestry Commission Forty-eighth Annual Report and Accounts 1966–1967 and the inclusion of the Comptroller and Auditor General's Report on the Accounts was noted either on the cover or the first page of the Report.

From the 49th CAR the year under report changed to the Government financial year ended 31st March.

This style of title continued until the 73rd Annual Report and Accounts 1992–1993, thereafter the number of the Report is omitted from the main title but appears on the spine.

From 1997–1998 Forest Enterprise Report and Accounts were published separately from the Forestry Commission Annual Report and Accounts.

Foreword

Britain can hardly be included among the great forestry countries such as Russia, Brazil or Canada. Despite a doubling in the size of its forests over the past 80 years, their area ranks towards the bottom of the country list in terms of hectares per person or percentage of forest cover. Nor can it be said that there exists in Britain a great long-standing forestry tradition or culture as found in Germany, France or Sweden, for example.

And yet .... there is a fascinating story to be told and achievements to be proud of in the history of British forestry during the 20th century. By the end of the century the area of forests in the UK (including Northern Ireland) had risen to over 2.75 million hectares and covered 11% of the land area. Three quarters of them consisted of plantations. Few other countries—Ireland and Denmark are two—have achieved a comparable change in the rural landscape in favour of forestry over as short a time. Furthermore, from being in a deplorable state by the end of the First World War, British forests are now well above the European average in terms of productivity (wood yield per hectare), and they are being called upon to meet increasingly heavy social and environmental demands from a dense, largely urbanised society. British forestry, and the institutions which support it, have come to be highly regarded by foresters worldwide.

Who better than E. G. 'Dick' Richards to tell this story? He gained his B. Sc. in forestry at Aberdeen University in 1938, but his career in forestry was delayed by the outbreak of the Second World War. He joined the British Army and saw active and distinguished service as an officer in the Royal Engineers in North Africa, Sicily and Western Europe. After the war, Dick joined the Forestry Commission as District Forest Officer in North Scotland and stayed with the Commission for the whole of his professional career until his retirement as Director at Headquarters in 1978. He held a number of positions, including responsibilities for the development of markets and the utilisation of timber, both from the Commission's forests as well as from the private sector. For many years he was a member of UK delegations to international meetings, including those of the FAO European Forestry Commission and the ECE Timber Committee, and he was also the Chairman of the Joint FAO/ECE/ILO Committee on Forest Working Techniques and the Training of Forest Workers.

Since his retirement from the Forestry Commission, Dick has remained active as a consultant in the field of forestry and forest industries. Especially during his more than thirty years with the Forestry Commission, Dick has been a first-hand witness—and contributor—to the evolution of forest policy in the UK.

Until 1919, the only forest policy was really a non-policy, an acceptance that as a great trading nation Britain would rely principally for its forest products needs, which were considerable, on overseas suppliers, notably countries in the northern coniferous belt and to a lesser extent the tropics. It took the catastrophe of the First World War, and in particular the shipping blockade, together with the felling of much of its remaining woodlands, for the realisation finally to sink in of its vulnerability in times of war. Thus by the Forestry Act of 1919 the Forestry Commission was established with the principal mandate to create a strategic reserve of timber. In the following 20 years considerable progress was made towards achieving this objective. By the time of the outbreak of the Second World War in 1939, 200 000 hectares of plantations had been established, the larger part by the Commission itself and mostly with fast-growing, exotic conifers, such as Sitka spruce. Because of their age these could contribute in only a minor way to supplying timber needs during and immediately after the war, so that the remaining mature woodlands, mainly in private hands, bore the brunt during those years, exacerbating the destruction wrought in the First World War.

British foresters found themselves on a steep learning curve in the inter-war years and after the Second World War war and benefited greatly from experience in other countries. Dick Richards' book is valuable in drawing attention to the role of the British Empire Forestry Conferences (later to become the Commonwealth Forestry Conferences), the first of which was held in the UK in 1920, followed by others in Canada (1923), Australia and New Zealand (1928) and South Africa (1935) and again in the UK in 1947. Since then further Conferences have been held at regular intervals, with the latest (sixteenth) in Australia in 2001. It is probably fair to say that British forestry gained a great deal from the longer-term experience of foresters in Canada, India and other members of what eventually became the Commonwealth. Increasingly, however, after the Second World War, the UK became involved in other international forestry activities, notably those of the Food and Agriculture Organisation (FAO) and its European Forestry Commission and the UN Economic Commission for Europe's Timber Committee. Its accession in 1973 to the European Economic Community, later the European Union, further strengthened its collaboration in forestry with other European countries, and as it built up expertise in many specific fields, it was able to contribute more and more to the exchange of international knowledge as well as to benefit from it. Still more recently, the UK has played an active role in the various follow-up activities to the United Nations Conference on Environment and Development in Rio de Janeiro in 1992 and the Strasbourg, Helsinki and Lisbon Ministerial Conferences on the Protection of Forests in Europe. Forestry in Britain truly came of age in the latter part of the 20th century.

By chronicling the evolution of forestry in the UK decade by decade, the book shows how official attitudes were translated through policies into actual measures. For example, in the decades after the Second World War, the common belief was that worldwide availability of wood would tighten over the long term, that Europe could not meet its growing demand for forest products and would become increasingly dependent on supplies from overseas, and that prices would rise. In those circumstances, it seemed logical for the UK to pursue its pre-war policy of expanding its domestic timber reserves and reducing its dependence on imports. However, an important change in post-war policy was to more actively involve the private sector to achieve this goal, and consequently grants, tax concessions and technical support by the Forestry Commission were made available to private forest owners. The result was that the post-war expansion of the forest estate was shared between the State, i.e. the Forestry Commission, and private owners. Another shift on the part of the Forestry Commission was to give increasing attention to the marketing of wood, mainly small-sized, that began to become available from its plantations, including the modernisation of the sawmilling sector and establishment of new industries, such as wood-based panels, pulp and paper.

During the last quarter of the 20th century, significant shifts occurred in Britain's forest policy, which were in response to external as well as internal stimuli, and which are described in the book. Not necessarily in chronological order, some of the most important ones were:

- While continuing to pursue the objective of building up the national timber reserve, the emphasis switched from State (Forestry Commission) forests to private ones by support for the latter through grants and subsidies and, until there was a change in legislation, tax concessions. Latterly, most afforestation was being carried out by private owners, and today more than half the area of plantations are privately owned and managed:
- As it became increasingly apparent from international studies that the expected global shortfall in wood supply was unlikely to happen and that in Europe forest growth was higher than the volume being harvested, the justification for continuing to invest heavily in building up British timber reserves at whatever cost needed to be finessed by finding additional arguments for supporting forestry in Britain. These included the contribution it could make to rural employment and to preserving the social infrastructure in the countryside, as well as the provision of non-wood goods and services, notably recreation.
- In conformity with the political philosophy of the Government at the time, part of the Forestry Commission's forest estate was sold in the late 1980s and early 1990s to the private sector. However, in face of unexpectedly strong opposition to this scheme from many quarters, further 'privatisation' was stopped in the mid–1990s;
- The original single-objective strategy—to build up timber reserves—gradually evolved towards a multi-objective policy in face of the public's increasing demand for recreation and environmental services, notably nature protection;

- Allied to this, forest planning and management had to show greater sensitivity towards the landscape, initially aroused by public objections to the planting of coniferous monocultures in rectangular blocks on hillsides, and to seek to be integrated better into overall rural—and increasingly urban—development planning;
- Management and silviculture also began to adapt to the exigences of sustainability and biological diversity, as espoused at UNCED and other international fora by, amongst other things, introducing a wider variety of species into forest stands, in particular broadleaved species. By the end of the 1990s, the area of broadleaves being planted annually in Britain exceeded that of conifers, in marked contrast to the trends in previous decades;
- The criteria used by the Treasury for the allocation of public funds to forestry, which had earlier been based on the expected financial return on the production of wood, increasingly made allowance for the provision of social and environmental services by forests, even though these could seldom be assessed in strictly monetary terms.

Perhaps more than in any other country, the questions of profitability and return on investment have played a central role in the forest policy debate in Britain. The background for this and the way the debate developed over the years are described in detail in the book and are one of the main reasons why it is so worthwhile reading. It is no secret that forestry has always had its opponents in certain segments of British society and in particular in some branches of officialdom, who sought to reduce Government financial support for forestry, which they considered to be an unnecessary and misdirected drain on public funds. Fortunately for British forestry, counter-arguments based particularly on its social and environmental roles have prevailed, notwithstanding some glitches along the way, for instance when changes in the tax laws in the 1980s brought private afforestation almost to a stop for a while. Such incidents demonstrated the vulnerability of a long-term activity such as forestry to short-term political whims; and conversely, the need for consistency in policy-making even while allowing for fine-tuning along the way.

Dick Richards leads us through the twists and turns of British forestry and forest policy over the past 80 years in a clear and entertaining way. He has been too modest to mention the contribution that he himself has made, as a past senior official of the Forestry Commission, to helping British forestry achieve the status that it enjoys internationally today. I hope that everyone who reads this book will enjoy it as much as I have done and will gain insight into the history of a fascinating and remarkable forest story, one which will continue to evolve in the 21st century.

Tim Peck
January 2002

Figure 1. Horse extraction of larger thinnings of Douglas fir, Eggesford Forest 1956

Author's Foreword

My aim in writing this book is to give an account of the creation of Britain's forest estate during the 20th Century.

Until 1919 Britain was one of the very few European countries that had no Government Forestry Policy and which had tolerated the depletion of its once abundant woodland cover to the point where it occupied only some 6% of the total land surface—one of the lowest in Europe where the average is around 25%.

The economic argument was that as a trading nation Britain's exports would enable her to buy her (very large) requirements of wood and wood products from the forest-rich countries of the northern coniferous belt (Canada, Scandinavia, the Baltic and Russia) and from the vast rain forests of the tropical countries, many of them part of the (then) British Empire.

Up until the First World War all went according to plan, but wood is a bulky commodity and importing the war-time essential timber requirements in 1914 and 1915 occupied between one-seventh and one-eighth of all the shipping entering British ports.

In 1919 The Forestry Commission was established by Act of Parliament with the general duty of promoting the interests of forestry, the development and expansion of Britain's forests and the production and supply of timber.

The adoption of a positive policy of direct tree planting by the State (the newly formed Forestry Commission) and the encouragement of the private sector to restock its woodlands, stripped of much of their mature timber for the war effort, did little to add to timber availability by the out-break of the Second World War some 20 years later in 1939. Again, the war-time needs for timber put a huge strain on shipping and further depleted the already low reserves of (mostly private) timber from home woodlands.

The Post-War Forestry Policy developed during the (1939–1946) war years and put into effect soon after the end of the war, saw a great increase in the area planted annually by the Forestry Commission and, in time, by the private sector, under the new long-term grant aided schemes for re-stocking and new planting.

The next phase of development began in the mid 1950s when the Forestry Commission set about encouraging the establishment of new forest products industries by private enterprise to utilise the rapidly increasing quantities of mainly coniferous thinnings from the new plantations; this was followed by the modernisation and expansion of the home-grown timber sawmilling industry as the plantations began to yield larger trees of sawlog size.

The emphasis is on the strategic targets for and achievements in creating productive woodlands rather than on the silvicultural means of achieving them; nor are details given of the industrial processes used in the forest products industries.

The economic arguments for and against State and private forestry—as opposed to other forms of land use (where these are practical alternatives)—are not discussed.

The internationally acclaimed work of the Research branch of the Forestry Commission, which covers forest mensuration as well as all aspects of silviculture merits a book in its own right. So too do the Commission's highly successful long-term efforts to improve operational efficiency including its continuing programme in the field of health, safety and training of forest workers and the mechanisation of forest work. A list of references covering these subjects is given at the end of this Foreword [**1–3**].

The main achievements from 1919 to the end of the Century may be summed up as follows:

- The forest area (State and private) increased from 6% of the total land area to some 10% (2.5 million hectares).
- The annual production of wood from British forests increased from 1.5 million m^3 (average 1934–1938) to over 9 million m^3 by the year 2000[1].

Capital investment in the wood processing industries in the 1980s and 1990s has been of the order of £1000 million. In spite of their being capital intensive and highly automated, they had over 11 000 employees.

The achievement in building up such a large area of forest with such a large (sustainable) annual harvest of wood is all the more remarkable when it is realised that most of the new plantations were planted on land quite unfit for arable agriculture and often unable to support viable sheep grazing even under the highly subsidised farming regimes of the post-war years.

The forest industries—sawmilling and the panel products (fibreboard and particle board) and paper pulp mills—had also to compete against the industries in forest-rich countries with huge quantities of relatively cheap wood from natural forests i.e. with no investment costs for the wood removed in the initial harvest and with low transport costs from forest to factory.

In large measure it was the daunting tasks faced by both growers and the wood processing industries and the need for them to co-operate closely that led eventually

[1] The contribution that Britain's forests have made to national timber needs is discussed in detail in Chapter 10.

to the establishment of a successful British forestry and forest products industry in its widest sense.

And this in spite of the loss of the once major outlet for mining timbers in the coal mines, when coal met most of industrial Britain's huge energy demand (the production of mining timber from plantations created under its 1919 Forestry Act policies was seen by Government as meeting a vital strategic need in times of war).

Forestry itself has changed out of all recognition. For the first half of the Century in Britain, virtually all operations were carried out manually. For example:

- In the nurseries where plants were raised from seeds, the garden spade, the fork, the hoe and the rake reigned supreme and the summer weeding of the acres of seedbeds was carried out by hand (this tedious job was given wherever possible to forestry students seeking "practical experience", to the latest joined members of the squad and to casual workers, including women earning so-called 'pin money').
- Ditching and cutting and the laying out of the turves in which trees were commonly planted and of the planting itself, were carried out by hand, often using specially designed spades and other tools.
- Tree felling was by hand saw and axe.
- Extraction from stump to roadside was by horse.
- Lorries were loaded manually, using hand powered wire cranes and winches for larger logs (the hydraulic crane had not then been designed).
- The worker wore little if any protective clothing; apart from rubber boots no working clothes were provided.

The 1950s saw a gradual change towards mechanised forest operations, a change which followed that in agriculture, but which was slower due to the need to develop much more specialised tractors and ploughs than were being used in farming and machines for harvesting trees, which met with altogether different problems from those found in harvesting grain and other farm crops, such as much steeper and more dangerous terrain and the large size and weight of the individual tree trunk with its many branches, which must be removed at some stage.

Table 1 gives a broad indication of the changes in the Forestry Commission's main activities—creating forests and harvesting wood—and in its requirements for industrial workers (= forest workers) to achieve the different programmes. Changes in the numbers of non-industrial workers are affected by the demands put on the Forest Authority aspects of the Commission's responsibilities as well as by its own programmes of work; the table gives no guidance on the relevant influence of these two factors on the numbers of non-industrial workers employed.

In the 1950s and 1960s upwards of 12 000 forest workers were employed to establish some 24 000 hectares of woodlands annually and to harvest a quarter to half a million cubic metres of wood (part of the annual harvest was worked by timber merchants buying the timber standing in the woods).

Table 1 Changing productivity as measured by numbers employed by the Forestry Commission and corresponding work loads

Year	Number employed		Area under plantations ($\times 10^3$ ha)	Area planted[a] ($\times 10^3$ ha)	Volume timber produced ($\times 10^6 m^3$)	Area PW grant scheme ($\times 10^3$ ha)
1	2		3	4	5	6
75 CAR	NI	1972	863	—	4.3	822
1994–1995	I	2024				
	IC	2654				
	Total	6650				
70 CAR	NI	2069	898	11	3.5	832
1989–1990	I	2781				
	IC	2675				
	Total	7525				
60 CAR	NI	2179	891	17	2.1	529
1979–1980	I	5950				
	Total	8129				
50 CAR	NI	2617	698	24	1.4	399
1969–1970	I	7487				
	Total	10 104				
40 CAR	NI	2649	521	24	0.7	252
1958–1959	I	12 512				(1960)
	Total	15 161				
30 CAR	NI	1113	270	25	0.3	15
1948–1949	I	12 100				(1950)
	Total	13 213				

NI = Non-Industrial I = Industrial
IC – Industrial Contractors working in FC woods including those harvesting standing timber bought by timber merchants. The number of industrial contractors is not recorded prior to 1990. Although their use was less common in the preceding decades, the figures for industrial workers from 1949–1980 are understated.
CAR – Commissioners Annual Report to Parliament
[a]Average of 5 years from 1949, 1959, 1970, 1980 and 1990 – afforestation plus restocking.

Table 1 shows in broad terms the reduction in the industrial labour force over the 46 years from 1949 to 1995, due in part to the reduction in the annual planting programme (still a relatively labour intensive operation). The massive increase in timber production has been achieved with a reduced labour force due to the replacement of man and horse by harvesting machines.
The use of computers has made it possible to reduce the non-industrial work force—central administrators and field managers and supervisors.

In the 1970s a similar planting programme and a two-fold increase in the volume of timber harvested was achieved, with only 60% of the labour force of the 1950s and 1960s; while the area of plantations to be maintained had more than doubled since 1950 and road building in the forest had also increased substantially.

From 1970–1980 the labour force decreased by 20% with an increase in the area of plantations managed of 28%; a decrease of 20% in the area planted; and a 50% increase in the volume of timber harvested. The increasing use of contractors (rather than direct employees) makes these simple comparisons less meaningful than the figures for the earlier decades. However, the comparison between the productivity of individual operations in the 1990s and productivity for corresponding operations a few decades earlier is quite startling. At its most dramatic, the change corresponds to the harvesting of grain by combine harvester and the horse-drawn binder (reaper with sheaf tying facility).

Mechanised felling has become an everyday reality. An operator working in the air-conditioned cab of a tractor-mounted timber harvester controls a chain saw on the end of a hydraulically operated boom to fell the tree. The branches are removed by de-limbing knives mounted in the felling head.

The length and diameter of the log is measured by the inboard computer, which then guides the chain-saw to cross-cut the tree into exact lengths and diameters that have been pre-set to meet individual customers' requirements. Up to 500 trees a day can be felled and processed in this way.

A single timber harvester can produce around 400 tonnes a week; a man working with a chain-saw might produce some 25 to 30 tonnes a week.

For safety reasons each machine is contacted every hour by a radio signal which the operator has to cancel.

The large flotation tyres of timber harvesters exert less pressure on the forest floor than a human footstep.

Altogether, Forest Enterprise produces some 80% (1 million tonnes) of its annual programme by timber harvester.

There are still situations where the difficulty of extraction or the small scale of the operation means that horse-extraction is the most cost-effective. Depending on the size of timber being extracted and the terrain, the horses vary from heavy Shire horses, such as Clydesdales, through Ardennes and Percheron heavy horses to Cumbrian ponies.

When the supply of pulpwood to the new pulp mill near Fort William was being planned in the early 1960s, the pulp mill owners were keen to reduce the impact of the use of the Highland road network—never designed to carry industrial traffic on any scale—by the large fleet of lorries needed to carry wood to the mill over a radius of over 100 miles.

Although much use was made of empty coal wagons to carry pitprops to the coal mines at that time, there were no rail wagons in use by British Rail designed specifically to carry round timber in bulk. And some of the rolling stock that might have been adapted for this purpose had old fashioned braking systems which meant that it was ill adapted to run on lines also being used by fast modern trains. The company persisted, however, and eventually, due to high level negotiations, pulpwood trains were built from old rolling stock to use a section of the Glasgow to Fort William Highland line, which was then carrying only a few trains a day. A pulpwood storage and handling facility was established at Crianlarich where the transfer from road to rail saved a long journey on poor roads to Fort William.

This operation was logical in the poorly-roaded Highlands but over most of the rest of the country it was much more economic to deliver the timber direct from the forest road to the pulp and board mills by lorry.

The building of the motorway system reinforced the economic argument for delivery by road from forest direct to mill.

With time, as the production of timber from home woodlands increased, the capacity of individual processing plants increased and the mechanised handling of pulpwood became more efficient, the economics of taking timber from forest to a nearby railway station for transfer onto rail wagons designed for round timber carriage improved significantly.

Rail freight operators could see the advantage of entering a business with good long-term prospects for growth—as indicated by the Forestry Commission's published forecasts of a rapid increase in the output of wood from its own and private woodlands. Long distance timber carrying freight trains now run regularly to major wood processing plants, not only from the large forests of Scotland but also from stations in the North of England.

The Commission working with private shipping interests has also (in the 1990s) solved the problems of shipping timber by sea along the west coast of Scotland, thus again removing a large tonnage of traffic from the rural road network of that region.

In addition to the technological revolution in forestry working methods, two other changes of similar magnitude have occurred. First the management style of the Forestry Commission and second the public perception of forestry policy.

In common with countries with a much longer tradition of forestry as a productive use of land and a rural enterprise running alongside agriculture, in 1919, the Forestry Commission was regarded as and organised as a Forest Service. It was not, however, a uniformed service as one found in France, Germany, Italy and many other European countries, where some forest services had also civil duties—such as controlling poaching in the forest, and avalanche, earth slip and flood control.

In the post-Second World War period, the Commission was organised like many other forest services with the forest as the basic working unit under a senior forester with academic qualifications from a forester training school of similar standard to an ordinary degree in forestry but with more emphasis on practical work. Forests were grouped into districts under a district forest officer and districts into Conservancies (5 in England, 4 in Scotland and 2 in Wales). Each country had a Director of Forestry in its capital city, but this link in the command chain was eventually cut out and Conservators reported direct to Commission Headquarters.

A feature of this management structure was the importance attached to senior officers—including particularly the Chairman R.L. Robinson, later Lord Robinson (himself a distinguished forester)—carrying out formal inspections of field operations. These had three main objectives:

- To meet field staff at forest and district level, often working in remote isolated areas with little or no public transport.
- To spread information on the best practices;
- To carry out a technical audit of the work being done.

In this context it must be remembered that all staff were on a steep "learning curve" as new methods of establishing and thinning plantations, based on research branch and other experiments, were being evolved.

Detailed notes of all the Chairman's visits were made and stored in Headquarters. Whenever he went on tour his secretary provided him with a summary of all his observations and instructions he had issued on his last visits to the forests concerned; complaints made to him and the follow up by his staff; technical queries he had referred to Research Branch; and other relevant matters. Not realising the efficiency of his backup and pre-visit briefing system, field staff were often surprised by the Chairman's detailed memory of what had been seen and discussed some years previously.

Although in time the Commission streamlined its line of command from Headquarters to the Forester-in-Charge of the Forest, the working unit, it was considered important that the eleven Conservancies be retained, each big enough to be managed by a Conservator, were small enough for him to know his area and be known to the general public and to other organisations and interests connected with the land.

In recent years this has changed totally as the revolution in information technology, the modern road network and the efficiency of the modern motor car has made it possible to manage the Forestry Commission with a greatly reduced supervisory field staff.

With a much smaller number of forest workers (now frequently machine operators rather than manual workers), the Commission presence in the countryside is altogether much reduced in the same way as is that of the farm worker. To many communities and sometimes to the public at large, this means that forestry has become identified with the commercial exploitation of the forest by an organisation interested only in making money—a view which is strongly held, if incorrect—and which in part accounts for the desire of individuals and organisations to have a say in the formulation of forestry policy and to manage local Forestry Commission forests as community forests.

At the same time, greater efficiency in carrying out forest operations and managing the Commission's forests has been claimed following the recent complete separation of the Forest Enterprise as manager of the Commission's forests from the Forestry Authority, which carries out the regulatory functions of the Commission and implements the grant-aid schemes to private growers, approves the Forest Enterprise programmes from the environmental point of view and represents the UK at international level.

In passing it may be noted that this move has been seen by some as a move to maximise revenue at the expense of other objectives such as those enshrined in the Wildlife & Countryside (Amendment) Act 1985 which placed a new duty on the Forestry Commission to endeavour, in carrying out its functions, to achieve a reasonable balance between the interests of forestry and those of the environment. Suspicions that Forest Services pursue a goal of maximising sustainable timber production is a situation not unique to Britain. The so-called "green" lobby in many European countries sees their Forest Service or Forestry Department as the government agency responsible for carrying out policies which have been formulated not by the Forestry Department but by a broad spectrum of government departments which give greater weight to nature conservation and amenity.

The reaction in Britain to such criticisms has been for the Forestry ministers and the Commission to produce greener and greener policy statements, some with barely a mention of the wealth generated by production of wood on a sustainable basis without which there would be no money to sustain the non-wood functions and benefits of the forests. One hopes, however, that in practice the reality is very much along the lines of the Commission's 1986 Forestry Facts (Booklet 1) which succinctly sums up its legal status and gives a well balanced account of how its different objectives will be achieved; and the much later 1994–1995 corporate agenda of the Forest Enterprise.

The following extracts from the Forestry Facts Booklet (1986) are relevant:

The Commission has both the status of a Government Department and a state-run industry. In its Ministry role, it reports directly to three Forestry Ministers—the Secretary of State for Scotland (the 'lead' Minister), the Minister of Agriculture, Fisheries and Food in England and the Secretary of State for Wales. It is responsible for giving these three Ministers advice on forestry policy and then for implementing any policy that may be decided by Government.

The Commission is, however, different from the usual Department of State in that it has a statutorily appointed Chairman and Board of Commissioners whose duties and powers are prescribed by law.

The Commission's responsibilities do not extend to Northern Ireland.

The Commission's Functions

The Commission has a dual role. It is simultaneously a trading concern responsible for the management of the nation's woodland estate (the Forestry Enterprise (a)) and the adviser on and executor of Government forestry policy (the Forestry Authority (b)). The Forestry Authority's functions include research, plant health (for trees and timber), tree felling control, grant aid for the private sector and representing the UK on forestry matters internationally, especially in the European Community. The Commission's approach to these distinct but complementary roles is reflected in its objectives.

(a) The Forestry Enterprise was renamed Forest Enterprise in 1992.
(b) The title Forestry Authority was dropped in favour of Forestry Commission in year 2000.
(c) The devolution of forestry to the Scottish Parliament and the Welsh Assembly, which came into effect at the turn of the this century, is not within the scope of this book.

The Commission's Objectives

Briefly stated, the Commission's objectives reflect the two functions described above.

As Forestry Authority its objectives are:

- To advance knowledge and understanding of forestry and trees in the countryside.
- To develop and ensure the best use of the country's forest resources; and to promote the development of the wood-using industry and its efficiency.
- To endeavour to achieve a reasonable balance between the interests of forestry and those of the environment.
- To undertake research relevant to the needs of forestry.
- To combat forest and tree pests and diseases.
- To advise and assist with safety and training in forestry.
- To encourage good forestry practice in private woodlands through advice and schemes of financial assistance and by controls on felling.

As Forest Enterprise, the Commission's objectives are:

- To develop its forests for the production of wood for industry by extending and improving the forest estate.
- To manage its estate economically and efficiently and to account for its activities to Ministers and Parliament.
- To protect and enhance the environment.
- To provide recreational facilities.
- To stimulate and support employment and the local economy in rural areas by the development of forests, including the establishment of new plantations and of the wood-using industry.
- To foster a harmonious relationship between forestry and agriculture.

The Corporate Agenda of Forest Enterprise 1994–1995

The Corporate Agenda of Forest Enterprise 1994–1995 gives in detail its corporate aims and objectives and the means by which it hopes to achieve them. It paints a re-assuring picture as to how it will balance on the one hand its main task, the efficient production of wood on a sustainable basis, with on the other its objectives of delivering, in such varying amounts as the individual forests may allow, the non-wood benefits of the forest in the fields of nature conservation, amenity and public recreation.

The following brief extract from the Corporate Agenda of Forest Enterprise 1994–1995 succinctly sums up the Forestry Commission's philosophy for its management of the State's forests into the next century.

"In accordance with Government forestry policy and under the direction of the Forestry Commissioners, Forest Enterprise is responsible for the management of the national forest estate containing some 875 000 hectares of forest and woodland. This is almost 40% of the area of forest in Britain".

Extracts from the Chief Executive, Forest Enterprise Introduction to its Corporate Agenda

"The meaning of sustainability in forestry refers not only to wood production, but to the sustainability of forest ecosystems, to the sustainability of landscape and visual amenity and to the sustainability of public enjoyment and recreational values.

This principle is embodied in the two main aims of the Government's forestry policy, namely, the sustainable management of our existing woods and forests and a steady expansion of tree cover to increase the many, diverse benefits that forests provide.

Multi-purpose forest management is the main instrument of the Government's Sustainable Forestry programme. In it, we aim to strike a balance between the production and service functions of forestry according to the opportunities and circumstances of every forest and woodland.

- Forest Enterprise manages the national forests in accordance with Forest Design Plans which fuse together wood production, environmental and recreational objectives;
- our design plans demonstrate sustainability through a long-term approach to management;
- our management guidelines safeguard and enhance biodiversity. We protect rare and endangered species and we take active measures to preserve and expand the remnants of the natural forest;
- forest operations are carried out according to an environmental code;
- public involvement in our management is encouraged through our Forest District Environment Panels and local authorities are consulted on each formal review of our Forest Design Plans."

Crofter forestry has been given a chapter on its own on two counts; the background to the failure of past initiatives to try to stimulate tree planting on Scottish crofts is not widely known; and the Crofter Forestry (Scotland) Act of 1991 has heralded in an unprecedented interest in and practical action in establishing forestry schemes in the Crofting Counties of Scotland.

The success or otherwise of a positive forestry policy can be judged in the most simplistic, but nonetheless telling way, by looking at the increase in the production of wood that has resulted from that policy. In the post-war years the increasing proportion of Britain's total consumption of wood and wood products that was coming from Britain's woodlands, was often quoted in Parliament as signalling the success of Government forestry policy. The published figures for the production of wood from Britain's forests are analysed in Chapter 10. The Food and Agriculture's Forestry Division figures and the similar but slightly different data published for the Commonwealth Forestry Conferences by the Forestry Commission between 1950 and 1960 are both given, because, confusingly, sometimes one set of figures and sometimes the other is quoted in literature without the source being identified.

From 1965 onwards the latest revised FAO figures are given.

These are recognised as the authoritative series of comparable data. They are the figures now used by the Commonwealth Forestry Conferences which have not undertaken their own calculations since 1962.

The figures show that the average Great Britain production of wood was something just under 1.5 million cubic metres a year from 1934–1938 and met about 4% of (UK) consumption.

By 1990, production (UK) was some 6.4 million m^3, meeting 13% of the increased UK consumption. It was forecast to double this figure by the year 2010. It had risen to just over 7.5 million m^3 by 1995 and over 9 million m^3 in the year 2000.

I have thought it appropriate to include, in summary form, the proceedings of the 15 Commonwealth Forestry Conferences (British Empire Forestry Conferences up to 1947) because of their strong influence on the formulation of British Forestry Policy in the early years and their continuing role as an important international forum with influence on the forest policies of the many very diverse countries that make up the Commonwealth Forestry Conferences of the latter half of the 20th Century and are set to continue into the 21st Century.

Throughout most of this book I have concentrated on reporting facts, giving references wherever possible to published sources. Only occasionally have I made a comment based on personal opinion and I have made it clear to the reader where I have done so.

I am aware that this factual approach may not always explain why British forest policy and practices, although influenced by other European, including Nordic, countries sometimes differed from them, nor does it explore the extent to which British forestry policies were influenced by international bodies like FAO and, in the closing decades of the Century, by British membership of the European Union.

So I have added comments on:

- The first 50 years of British forest policy (1919–1969).
- The period 1970–1980.
- The period 1981–2000.

The influence of the Empire and Commonwealth Forestry Conferences is dealt with in some detail in the Introduction and in Chapter 12.

My personal comments are based on 31 years service with the Forestry Commission, from 1946 to 1978, of which some 21 years were spent at Commission Headquarters when post-war policies were being formulated and the Commission as forest authority began to make regular, increasingly frequent, contacts with other European and Scandinavian countries and international forestry organisations, in which I was much involved.

Forestry Commission Senior Staff

The Commission has been under the guidance of only 12 chairmen since 1919. The longest serving chairman, Lord Robinson, was the key member of the Commission staff from its inception in 1919 and was the leading light in the immensely important period after the Second World War; he was also a key figure in Commonwealth forestry.

In the period since 1946, when the post was created, there have been only 9 Director Generals, all professional foresters, except Sir William L. Taylor who was a professional land agent, with training in forestry.

Forestry Commission Chairmen

1919–1927	Lord Lovat
1927–1930	Lord Clinton
1930–1932	Sir John Stirling Maxwell
1932–1952	Lord Robinson
1952–1963	Lord Radnor
1963–1965	Lord Waldegrave
1965–1971	Leslie A.W. Jenkins
1971–1976	Lord Taylor
1976–1979	John Mackie, later Lord John Mackie
1979–1989	Sir David Montgomery
1989–1994	Sir Raymond Johnstone
1994–2001	Sir Peter Hutchinson

Forestry Commission Directors General

1946–1947	Lord Robinson
1947–1948	Sir William L. Taylor
1948–1962	Sir Arthur H.L. Gosling
1962–1968	Sir Henry Beresford-Peirse
1968–1976	J.A. Dickson
1976–1986	G.D. Holmes
1986–1990	G. Francis
1990–1995	T.R. Cutler
1995–	David Bills

References

1. W.O. Wittering (Editor). *Work Study in Forestry.* Forestry Commission Bulletin 47. HMSO 1973.
2. R.F. Wood. *Fifty Years of Forestry Research.* Forestry Commission Bulletin 50. HMSO 1974.
3. Forestry Commission. *Report on Forest Research.* Published annually since 1950. HMSO.

Introduction

Historic Background

Following the end of the last Ice Age—about 10 000 years ago—trees began to re-establish themselves in Britain. By the time Neolithic man arrived, around 4000 B.C. much of the land was covered in forest, except for mountain tops, marshes and small areas of natural moorland grassland. In upland England and Wales and in much of Scotland, hazel–oak woodlands predominated; the commonest trees found in lowland England and Wales were lime, hazel, oak and elm. In the Scottish Highlands the woods were mainly of birch and Scots pine.

The process of clearing forests for agriculture and using the forests as grazing for domestic animals was the main cause of deforestation and by around 500 B.C. only about half of England remained wooded. The conversion of land to agriculture continued under the Romans and the Anglo-Saxons in England so that by the year 1000 A.D. the forest area had fallen to less than 20% of the land surface. Wood requirements were met from a few actively managed woodlands, particularly coppices.

As the population increased, so did the demand for agricultural land and by the mid-14th century only about 10% of England remained as forest; Wales followed a similar trend. In Scotland much of the natural forest was cleared in the three centuries up to 1350 A.D. In line with what was happening elsewhere in Europe, the monasteries played a significant role in clearing forests e.g. for sheep grazing in Scotland and upland England.

Pressure on the forests eased somewhat after the decline in population caused by the Black Death (the Plague) in the period 1350–1550, but nevertheless the loss of woodlands became a matter of serious public concern. For example, restrictions were placed on the felling of wood for iron smelting in England in the 16th century. From the 17th century onwards, land owners began to take a great interest in growing timber as a commercial crop, notably to meet the growing demand for oak for ship building, which by the end of the 18th century had reached 100 000 m^3 per year. Plantations of Scots pine, European larch and Norway spruce were also established against future demand for timbers. Later North American conifers such as Douglas fir and Sitka spruce were introduced. But overall Britain came more and more to rely on imported timber and with some notable exceptions forestry as a commercial activity began to be of less interest, as many landowners started to develop sheep grazing in the uplands and the sporting potential of their estates. For example, sheep grazing and deer stalking on open moorland in Scotland became a major land use.

The Industrial Revolution and The First World War

When the Industrial Revolution created a rapidly increasing demand for cheap easily worked softwoods (conifers), home woodlands had neither the types nor qualities of timber to meet this demand. In England and Wales woodland owners were slow to respond to this new market. They continued for the most part to grow traditional broadleaves, often with sporting cover in mind. The woodland produce was destined mainly for estate use and for local markets such as farm buildings, fencing, hardwood pit props (in Wales), charcoal and firewood. In southern England hazel and chestnut coppices were common.

In Scotland with its much higher proportion of conifers than England and Wales, the total area of woods was still very small in relation to demand and competition, in land use terms, from sheep for the wool trade, grouse moors and (red) deer forests remained strong.

The Governments of the day were content with this state of affairs. As a trading nation with a large Empire, it was seen as sensible to rely on imports from the forest-rich countries with seemingly endless supplies of coniferous timbers such as Canada, Scandinavia, the Baltic countries and Russia; temperate hardwoods (broadleaved timbers) from central and southern Europe; and tropical timbers from the vast rain-forests circling the world. The products of British manufacture, so dominant in world trade would, it was agreed, readily pay for these imports so why engage in the long-term financially unrewarding business of growing timber.

There were of course notable exceptions and some landowners built up extensive areas of woodlands managed with the production of timber as a main—though often not the only–objective. And in 1909 following a series of Government Committees some money was invested in the Development Commissioners for afforestation schemes; but little was in fact achieved.

By 1914 imports of coniferous timber had risen to the point where they met 93% of requirements [1]. And although imports fell from 11.5 million tons in 1913 to 2.5 million tons in 1918, they imposed a great strain on the shipping that had to run the blockade of British ports. In the first two years of the war, timber occupied between one-seventh and one-eighth of the total shipping entering the country.

The 1919 Forestry Act

Home production rose from less than one million tons in 1913 to 4.5 million tons in 1918, necessitating the cutting down of 450 000 acres (182 000 hectares) of woodlands, mainly privately owned, out of a total of 3 million acres (1.2 million hectares) [1]. The importance the necessity some might say of having an adequate supply of timber growing in the country in times of war was brought home in no uncertain way. The Government accepted the recommendation of the Forestry Sub-Committee (the Acland Committee) of the Reconstruction Committee [2] that a National forestry authority be set up to promote the interests of forestry, the development of afforestation and the production and supply of timber in the UK (which then included all of Ireland).

The Authority, the Forestry Commission, was to consist of eight Forestry Commissioners to be appointed under Royal Warrant. One would be Chairman, not less than two would have special knowledge and experience of forestry in Scotland, and one at least would have scientific attainments and a technical knowledge of forestry. The Commission would be empowered to give grant aid to private woodland owners whose contribution to the production of timber in the UK was judged to be vital.

The Forestry Act 1919 "an Act for establishing a Forestry Commission for the UK and promoting afforestation and the production and supply of timber therein, and for purposes in connection therewith" received Royal Assent on the 19th August 1919 and came into effect on the 1st of September [3]. The Forestry Commissioners were appointed on 29th November. Following the first Commissioners meeting on the 7th December in London, a "token planting"—the heeling in of some beech and larch—was made that evening by Lord Clinton on his arrival by train at Eggesford, Devon, and a telegram giving this news was sent to greet Lord Lovat on his arrival by overnight train on the 8th December at Elgin [4, 5]. The official record states that the first tree was planted officially on 8th December at Eggesford. Lord Lovat planted the first tree in Scotland later that same day.

These were symbolic gestures designed to show the urgency felt by the Commissioners in getting on with the massive task entrusted to them

They were helped by the fact that a small Interim Forest Authority had been established, in anticipation of the necessary forestry legislation being passed, to carry out preliminary work and to ensure that an adequate supply of plants would be available for planting as soon as the war ended.

By the end of their first Forest Year (30th September 1920) the Commission [**5**] had acquired a staff, planted 1595 acres (645 ha) on land they had acquired, surveyed some 645 000 acres (261 000 ha) of potential planting land, started a programme of research and experiment on the rates of growth of conifers under different site conditions, and set up Forester Training Schools at the Forest of Dean (by extending the existing Training School for Woodmen), New Forest, Chopwell in County Durham, Beauly in Inverness-shire and Avondale in County Wicklow.

They had also established relationships with the Universities awarding forestry degrees—Aberdeen, Bangor, Cambridge, Edinburgh and Oxford—and the diploma awarding Colleges of Armstrong (Newcastle) and Dublin's Royal College of Science.

As provided for in the Act, Consultative Committees for England, Wales, Scotland and Ireland had been set up to advise the Commissioners on the performance of their duties. Members were to include representatives of the appropriate Board of Agriculture; persons with experience of forestry and woodland industries; representatives of labour; representatives of county councils and other bodies and societies interested in or promoting forestry and of woodland owners.

At the time of the 1919 Forestry Act, the British Empire was intact and the role of Empire forests in supplying timber to the UK had an important bearing on UK forestry policy. Contradicting the conventional view of dependence on oversees supplies, the Acland Committee [**2**] had made the following points in setting out the case for providing a much greater proportion of UK demand for wood and wood products from UK woodlands:

- Dependence on imported timber is a grave source of weakness in war.
- Even in peace time supplies of timber 'lie too much outside the Empire' (where there may be political instability, as in Russia following the Communist Revolution).
- Within the Empire e.g. Canada, there is a risk of over-exploiting the natural forest resources (a point made again in 1921 by the Forestry Commission in the wider context of over-exploitation of 'the world's virgin forests').

And so to keep under review the forest resources and forest policies of the Empire, the Commissioners were given power to make such inquiries as they might think necessary for the purpose of securing an adequate supply of timber in the UK and promoting the production of timber in His Majesty's Dominions.

First British Empire Forestry Conference

In July 1920, the Forestry Commission organised the first British Empire Forestry Conference in London [**6**]. The first inaugural meeting was held in the Guildhall, London on 7th July. Conference meetings lasted for seven days and field and other visits for 12 days. The proceedings, resolutions and summary of statements occupy some 304 pages.

In addition to the UK (the 'Mother Country') Dominion and Empire countries attending the conference were:

Australia, Canada, Newfoundland, New Zealand, India, South Africa.

Crown Colonies attending were:

Ceylon, East African Protectorate, Federated Malay States, Gold Coast, Nigeria, Nyasaland, Sierra Leone, Tanganyika Territory, Trinidad, Uganda.

Egypt and the Sudan also attended.

Some 33 statements were presented to the conference, including (in addition to those listed above) statements from Swaziland, Southern Rhodesia, Bahamas, Bermuda, British Guyana, British Honduras, Hong Kong, Jamaica, Leeward Islands, Malay Peninsula, Seychelles, Weihaiwei, Windward Islands.

After the welcome by the Lord Mayor of London and the Conference's response, Lord Lovat was invited by the Chairman of the Conference (Viscount Milner, Secretary of State for the Colonies) to deliver the opening address. That address contained some pungent criticism of the United Kingdom's lack of a forestry policy and its negative attitude to forestry in general.

The following extracts from the speech give some idea of the no-nonsense, robust way in which Lord Lovat gave his views.

Lord Lovat: My Lords, Ladies and Gentlemen: Before opening the discussion on the subject which has brought us together, I would like to say one word of personal explanation on behalf of my colleagues and myself about the action of the British Forestry Commission, in point of age and experience the junior Forestry Authority of the British Empire.

When the suggestion was made from Canada and India that it would be advisable to hold a British Empire Forestry Conference at the same time as the Empire Timber Exhibition, it was necessary for some forest authority to take the responsibility for issuing invitations and for making the necessary preparations. It seemed to us to be the moment to lay aside the role of the Cinderella of Empire Forest authorities and take up a part for which we knew our inexperience made us in no way qualified.

The representative nature of the meeting today shows that the decision taken was a right one. I am glad to be able to report that we have here today representatives from practically every portion of the Empire.

Importance of Forests[2]

I will open the discussion by asking you to consider firstly the importance of forests and the products of forests to the British Empire. Secondly, to consider the apathy of our race in all questions connected with forestry, our story of neglected opportunities in the past, and the necessity for stimulating an interest in forestry in the future.

[2]Several paragraphs have been omitted as they do not add to the thrust of the argument.

Thirdly and lastly, to look for a moment at the long lists of opportunities which community of purpose and action may open up to the constituent units of the Empire.

It is sufficient to say as regards the British Isles that the value of our timber imports in 1919 reached the colossal figure of £72 000 000, that they will probably approach £90 000 000 to £100 000 000 this year, that our timber imports require some 13% of the total British tonnage for transportation, that before the War 10 000 000 tons of timber and timber products were brought into this country, which represents a tonnage space equal to that required for the whole of the grain imports and greater than that required for all the other foodstuffs put together with cotton and wool thrown in as well.

The part which timber plays in the life of every civilised State is set out by John Evelyn in his Sylva A Discourse of Forest Trees (17th century) when he says:

> Since it is certain and admissible that all arts and artisans whatsoever must fail and cease if there were no timber and wood in a Nation (for he that shall take his pen and begin to set down what art, mystery or trade belonging in any way to human life could be maintained and exercised without wood will quickly find that I speak no paradox) I say, when this shall be well considered it would appear that we had better be without gold than without timber.

These are weighty words, as applicable today as they were 200 years ago, and if we accept them, as we must accept them, the question we have to ask ourselves is "Is our forest policy, is our forestry practice as a whole worthy of our Empire? Are we as unit of civilisation taking the necessary steps to preserve the great heritage of timber to which we have succeeded?"

Let me take Great Britain first—for criticism, like charity, should begin at home—we in Great Britain were the last European State to adopt a forest policy, with hesitating steps, and haltingly at 25 years' interval we followed that enlightened country, Turkey, and established a Forest Authority. Of all nations of Europe we have the smallest area of State forest; we have by far the least area of wood per head of population and with one exception are the most poorly wooded country in proportion to our area of land. We lead in one thing and one thing only—the number of Inquiries, Royal Commissions, Departmental Committees, which have examined and re-examined the forest situation in Great Britain, without, until 1919, doing anything effective to have it improved.

If we pass from the forest history of this country to the Dominions and Crown Colonies we see the attitude of indifference of the Mother Country continued, sometimes even after they had emancipated themselves from her control. France by 1830 had established an Algerian forest programme. It was not until 1855, under Lord Dalhousie, that India led the way and was the first of the Colonies to organise a forest service.

We shall hear, in the course of our discussion, the date of establishment and conditions of forest services in the Dominions and Crown Colonies, almost all of whom have taken action before the home country. Many of these forest authorities have done admirable work, but some are still conducted without sufficient funds and in certain cases no definite and permanent policy has been laid down.

If we take the case of our blood relation, the United States of America, we find an almost exact replica of our own policy of inaction. As early as 1799, possibly under French influence, the first forest legislation was attempted.

Up to 1890, nearly a century later, a comprehensive federal forestry policy was still under discussion. It was not until 1903 that a separate and un-trammelled forestry service was established, and like us, it was not until the devastation of the woods had brought it home to the people that effective action must be taken, that extended "reserves" were placed under State control.

If we glance for a moment at the forest history of Europe, we are struck by the fact that in the Middle Ages, whilst forestry laws on the Continent were being built up on national and constructive lines, our forest laws were based on injustice, cruelty and repression. The forests, or rather game preserves of William the Conqueror and his successors—some 68 or 69 in number—were established by the expropriation of the rights of individuals and in certain cases the destruction of villages and the wasting of whole countrysides. The Norman rulers' forest laws were regarded by rich and poor, cleric and laymen, as galling acts of Royal despotism and though repealed in part at the time of the Magna Carta and by subsequent enactments, were not finally abolished in their entirety until from 1760–1817.

It is a significant fact, whether arising from British dislike of forests or lack of appreciation of the possibilities of communal forest enterprise that the area occupied by the British during the French Wars of the Middle Ages is the only area in France where no communal forest exists, and this in a country where five million acres of communal forests are to be found in the departments untouched by British rule!

We are not forest preservers as they are in Switzerland, France, Germany and Belgium, but at heart forest vandals who look on all woodlands as their prey.

One hundred years ago the Landes, some two million acres in extent, was a barren waste, grazed by a poor and unhealthy type of sheep, aptly described by a local writer as swampy, fever ridden and desolate. This area, by intelligent co-operation between State, communes and individual owners has been turned from the poorest district in France into two of the richest departments of that rich country.

If we look at the other side of the picture:

In Great Britain, there are many millions of acres devoted to sport or carrying not more than one sheep to four or five acres. It is probably true to say that there are approximately four to five million acres in the British Isles which are either incapable of producing or are not actually producing more than two to three pounds of mutton and rather less than one pound of wool per acre per annum. Probably not more than one-third to one-fourth of the total grazing area is suitable to tree growth, but admitting this and admitting also that if the low ground were planted, the high ground would be thrown out of gear and only available for summer grazing, it is certain that an extension of the planting area would greatly benefit the general community.

The benefit to the community of planting second-rate grazing is still greatly in favour of forestry. Mr. Acland's Reconstruction Sub-Committee calculated that when the whole of the additional 1 770 000 acres were afforested, which were necessary to afforest to secure this country against timber famine in case of a European War, the meat yield of Great Britain and Ireland would only be reduced by less than 0.7 of 1%.

It is not my intention today to make alarmist statements, or to quote panic figures. Without over-stating one's case, one can say without fear of contradiction, that the world's timber situation gives grave reason for thought and enquiry. We are informed by the leading periodicals and other authorities of a world's paper shortage; we read in the report of a recent American Commission that timber is being cut in the United States at three times the rate of growth, that a shortage has begun and will become increasingly acute for several decades. We know that several exporting States admit that they have over-estimated their forest resources. Whole groups of countries have been shut out of the timber trade by revolutions, dynastic changes and the rise and fall in the value of exchange.

Even if we do not accept the idea of a world timber famine, we must admit that never was there a time before in which the Empire depended more on its own efforts for its supply of timber, nor has there ever been a time more favourable than the present for urging the examination of our Empire Resources and, if thought advisable, the definition of an Empire Forest Policy.

I would submit that our first object—which we cannot hope to achieve in its entirety, but towards which we should make considerable progress at this Conference, is the removal of the question of Empire timber resources out of the realms of fancy into the land of bedrock fact.

Forest Authority

Another group of subjects which might lead to a fruitful exchange of ideas is the question of State Forest Policy; the functions of its Forest Authority and the Forest Authority's relations to other departments.

After many years of travail, the British Government in August 1919 gave simultaneous birth to a Forest Policy and a Forest Authority. I do not propose to enter into the details of the Forestry Bill of August, 1919, but I think some of the principles on which it is based are worthy of consideration and have been favourably reviewed by Continental Nations. It was decided:

1. That there should be one Forest Authority for Great Britain and Ireland.
2. That this Forest Authority should have the control of policy, finance, personnel,

education and research.

3. That the executive should be decentralised in England and Wales, Scotland and Ireland.
4. That a fixed Forestry Policy should be laid down for an extended period and that a block grant should be voted for a shorter period (10 years) with certain limited objectives to be carried out during that period.
5. That the control of Parliament over finance should be exercised by the annual vote of estimates.
6. That the position of forest officers should be secured by including them in the Civil Service after a period of probation.
7. That consultative committees should be appointed to represent the various bodies interested in Forestry in England, Scotland, Ireland and Wales.

These decisions give certain very definite advantages to the British Forestry Authority and *mutatis mutandis* are worthy of consideration for wider application.

The rest of the speech dealt briefly with the need for forestry education and for forest research.

The Conference passed a number of resolutions which have had a profound influence on UK forestry and on that of the Commonwealth of countries which grew out of the former Empire. The Conference still meets at regular intervals (as the Commonwealth Forestry Conference, CFC). The 15th meeting took place in Zimbabwe in 1997. Not least of its roles in the immediate post-Second World War years was to provide data to the Forestry Division of the Food and Agriculture Organisation (FAO) of the United Nations, based in Rome. Among other things the CFC had started to build up over the years (up to and including the second world war) valuable statistics on the forests of its member countries and on trade in forest products within the Commonwealth and between the Commonwealth and the rest of the world.

The relevant resolutions of the first Conference are given below.

Resolutions Passed at the Meeting Held on 22nd July, 1920

The British Empire Forestry Conference, consisting of the delegates from the Empire whose names are given herewith, assembled in London in July 1920, and sat in Committee on July 12th, 13th 14th, 20th and 21st and in full session on July 7th and 22nd. After deliberation and discussion they have passed the following resolutions which they will bring to the notice of their respective Governments.

Forest Policy

In view of the great importance to the Empire as a whole, as well as to each of its component parts, of producing a sustained yield of all classes of timber and of encouraging the most economical utilisation of timber and other forest products and of maintaining and improving climatic conditions in the interests of agriculture and water supply, each of the Governments of the Empire should lay down a definite forest policy to be administered by a properly constituted and adequate forest service.

Survey of Resources

The foundation of a stable forest policy for the Empire and for its component parts must be the collection, co-ordination and dissemination of facts as to the existing state of the forests and the current and prospective demands on them.

Constitution and Status

In order to attain continuity in the development of forest resources, it is desirable that certain elements of stability be secured in the constitution of the forest policy. This may be done by the following measures:

1. The definition, where this has not been done already, of a forest policy in a Forestry Act or Ordinance.
2. The reservation for the purpose of economic management and development of forest land under conditions which prevent the alienation of any which is primarily suitable for forests except for reasons consistent with the maintenance of the forest policy as a whole.
3. The assurance to the Forest Authority of funds sufficient to carry out the accepted policy for a series of years.
4. The grant to members of the forestry service of the status of civil servants with due provision for pension.
5. The appointment as the chief officers of the forestry service of persons having a high standard of training in forestry, their selection and promotion being by merit alone.
6. The establishment in each of the larger parts of the Empire and for the Colonies not possessing responsible government collectively, of an officer or officers, having special duties of advising as to forest policy and surveying its execution.

Organisation of Forest Industries

It is extremely desirable that the Forest Authority should be in close touch and consultation with organisations representing the interests concerned in the extraction and utilisation of timber and other forest products.

Publicity

It is the duty of the Forest Authority in every part of the Empire to adopt and encourage methods of education and publicity in order that the people may be fully informed of the aims and purposes of forest policy and may thus be induced to co-operate towards its successful fulfilment.

Resolutions 6–9 dealt with plant supply, terminology, research and education.

Forestry Bureau

The Conference approve the suggestions and recommendations for the constitution of an Imperial Forestry Bureau and strongly urge upon their respective Governments that they should contribute to the support of the Bureau. They feel that it will be largely upon the work of such a Bureau that the proper development of the forestry resources of the Empire will depend and they therefore cannot overemphasise its importance as a part of Empire organisation.

Future Conferences

The Conference is convinced that the holding of Conferences of Representatives of the Empire on forestry matters is of great service. They desire to thank the Forestry Commission of the UK for causing the Conference to be assembled and for making the necessary arrangements.

They recommend that the next Conference be held in the year 1923 and that, if the Dominion Government approves, it be convened in Canada.

May It Please Your Majesty:

On the 21st July, the Conference was received by His Majesty the King George V at Buckingham Palace. The address and the reply included the following:

We, the delegates to the British Empire Forestry Conference from the UK, the British Dominions beyond the seas and the Empire of India have the honour to present to your Majesty this humble address.

We value very highly your Majesty's gracious permission to be present here today, both because it enables us to submit this expression of our loyalty to your Majesty's Person and Throne and also as a recognition of the important position occupied by forestry in the resources of the Empire.

The Governments of the Empire have responded to the invitation from the Mother Country by appointing delegates and by causing to be prepared statements upon which we have based our deliberations describing their timber resources and the methods adopted for their conservation and utilisation and giving estimates of production and consumption which will be a guide to future policy.

We were fortunate in being able to visit the Empire Timber Exhibition. We believe that this Exhibition has greatly assisted Your Majesty's subjects to realise the supreme importance of a steady increase in the production of timber from all parts of the Empire, sufficient in quantity and quality to meet the ever-growing needs of the world.

Our work is still continuing and we cannot yet foresee its full effects. We trust, however, that there may result from it the setting up of an Imperial Forestry Bureau which may act as a centre of information with regard to all Imperial Forestry problems to which forest officers and foresters from all parts of the Empire may look for any assistance which they may desire to obtain.

Of one thing we have had abundant proof—that in forestry matters the different parts of the Empire in spite of differences of situation, resources and experience, have much to learn from and to teach one another. We trust that it may be found that this spirit of friendly co-operation which our Conference has evoked, inspired as it is by loyalty and devotion to Your Majesty's service may inure to the advantage of Your Empire for many generations to come.

His Majesty's Reply:

It gives me much pleasure to welcome you here today and I appreciate your reference to me as a brother forester. Both my father and Queen Victoria took a keen personal interest in forestry and I am glad to think that, partly owing to that interest, the Crown woods were able to make some contribution towards the needs of our armies in France and Flanders. I extend a special welcome to those of you who have travelled far from overseas to take part in the Conference. During the War, I came into contact with many of the foresters from overseas, including the Canadian Forestry Corps who were working in my Windsor woods and I know how valuable were the services they rendered.

I congratulate the Forest Authority of the Home Country that it has been able thus early to join hands with foresters in other parts of my Empire.

The work of the Universities and Colleges and the experience gained in the Crown woods and private plantations, have laid a foundation upon which it is incumbent upon us to build; and it is satisfactory to find that India and the Dominions, where forest science and administration already form part of the national life, are willing to place their experience at our disposal.

Their help will be most valuable to those parts of the Empire to which the importance of the subject has for the first time been seriously brought home by the experience of the War.

It is a peculiar difficulty of your work that it demands perhaps more imagination, more patience and more foresight than any other industry and it is an immense advantage that the experience of all parts of the Empire should be brought into a common stock and made available for all.

List of Delegates who Attended the Conference

United Kingdom

Major-General Lord Lovat, K.T., K.C.M.G., D.S.O., Chairman of the Forestry Commission.

The Rt. Hon. F.D. Acland, M.P., Forestry Commissioner.

Mr. R.L. Robinson, O.B.E., B.A., B.Sc., Forestry Commissioner.
Professor Sir William Schlich, F.R.S., K.C.I.E.

England and Wales

Mr. Hugh Murray, C.I.E., C.B.E., Assistant Forestry Commissioner

Scotland

Mr. John Sutherland, C.B.E., Assistant Forestry Commissioner

Ireland

Mr. A.C. Forbes, O.B.E., Assistant Forestry Commissioner

Australia

Mr. C.E. Lane Poole, Conservator of Forests, West Australia
Mr. H.R. Mackay, Commissioner of Forests, Victoria

South Australia

The Hon. Edward Lucas, Agent-General

Victoria

The Hon. Sir Peter McBride, Agent-General

Queensland

The Hon. James McEwan Hunter, Agent-General

West Australia

The Hon. Sir J.D. Connolly, Agent-General

Tasmania

Mr. A.H. Ashbolt, Agent-General

Canada

Mr. E.H. Finlayson, Forestry Branch of the Department of the Interior, Ottawa
Mr. Robson Black, Secretary, Canadian Forestry Association
Mr. C. Leavitt, Chief Forester, Commission of Conservation, Ottawa

Mr. Ellwood Wilson, Chief Forester, Laurentide Paper Company, Quebec
Mr. W.H. Kilby, Canadian National Railways

British Columbia

Mr. M.A. Grainger, Chief Forester, Provincial Forest Service

Quebec

Mr. A. Bedard, Assistant Chief Forester, Provincial Forest Service

India

The following representatives of the Government of India:

Sir Claude Hill, K.C.S.I., C.I.E., Member of Viceroy's Executive Council
Mr. P.H. Clutterbuck, C.I.E., C.B.E., Chief Conservator for United Provinces
Mr. A.J. Gibson, Conservator of Forests
Mr. W.F. Perree, C.I.E., President, Imperial Forest Research Institute, Dehra Dun
Professor R.S. Troup, C.I.E., formerly of the Indian Forest Service

Newfoundland

Sir Mayson M. Beeton, K.B.E., Newfoundland Development Company
Mr. D. James Davies, B.Sc., Department of Agriculture, Newfoundland

New Zealand

The Hon. Sir Thomas Mackenzie, K.C.M.G., High Commissioner

Union of South Africa

Mr. C.E. Legat, Chief Conservator of Forests

Crown Colonies

Colonial Office

Mr. W.D. Ellis, C.M.G., M.A.

Ceylon

Mr. J.D. Sargent, Deputy Conservator of Forests

East African Protectorate

Mr. Edward Battiscombe, Conservator of Forests

Federated Malay States

Mr. G.E. Cubitt, Conservator of Forests

Gold Coast

Mr. N.C. McLeod, Conservator of Forests

Nigeria

Mr. H.M. Thompson, C.M.G., Director of Forests

Nyasaland

Mr. J.M. Purves, Chief Forest Officer

Sierra Leone

Mr. L. Palfreman, Conservator of Forests

Tanqanyika Territory

Mr. D.K.S. Grant, Conservator of Forests

Trinidad

Mr. C.S. Rogers, Conservator of Forests

Uganda

Mr. R. Fyffe, Chief Forest Officer

Egypt and the Sudan

Mr. W.A. Davie, Sudan Agricultural Department
Major R.E. Fischer, Sudan Agricultural Department

Secretary,
Mr. O.J. Sangar, M.C.
Assistant Secretary,
Miss G. Baker-King

References

1. Forestry Commission. *Report of the Committee on Marketing of Woodland Produce* 1956 H.M.S.O. (Known as the Watson Committee)
2. Ministry of Reconstruction, Reconstruction Committee Forestry Sub-Committee. *Final Report Cd 8881.* 1918
3. *The Forestry Act 1919*
4. D. Pringle. *The Forestry Commission. The First 75 Years* (1994)
5. 1st CAR (*First Annual Report f the Forestry Commissioners.* Year ending September 30th 1920. HMSO)
6. *Forestry Commission (London) British Empire Forestry Conference, London 1920* Proceedings, Resolutions and Summary of Statements HMSO 1921

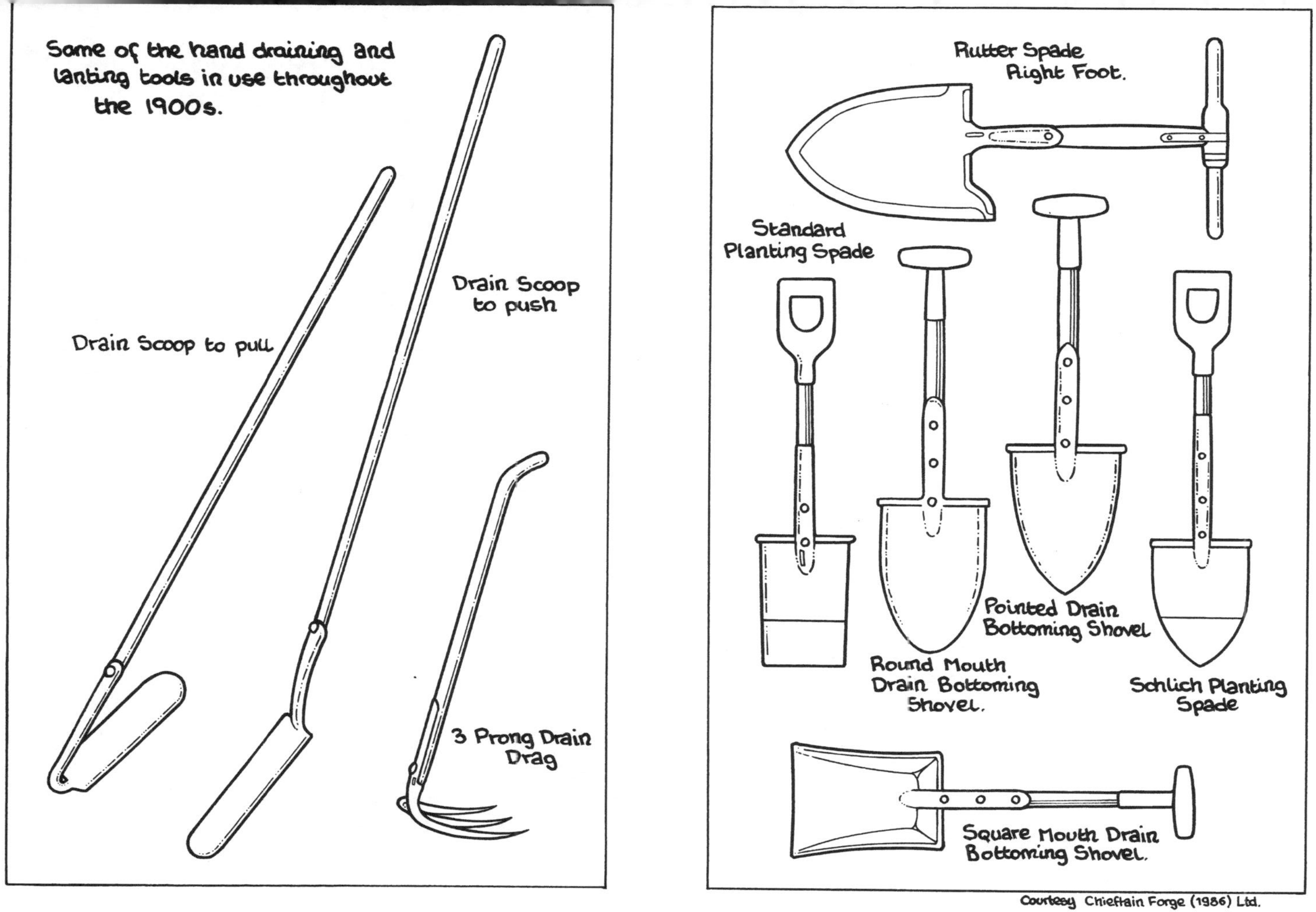

Figure 2. Some of the hand draining and planting tools in use throughout the 1990s

Chapter 1

The Inter-War Years 1919–1939

1 Selected Highlights

Before looking at the achievements and failures of the new Forest Policy in the inter-war years, two changes affecting the Forestry Commission's responsibilities must be noted. In 1922 forestry in Ireland came under the Government of the Irish Free State and of the provincial administration in Northern Ireland. The Forestry Commissioners responsibilities were confined to Great Britain. In 1923 most of the Crown Woods were transferred from the (Crown) Commissioners of Woods and Forests [**1**]. The 120 000 acres transferred included the New Forest and the Forest of Dean, Bedgebury (Hants) Delamere Woods (Chester), Inverliever (Argyll), Hazelborough Woods (Northants), Dymock Woods (Gloucester) and Chopwell Woods (Durham).

Between 1919 and 1939 the Commission had built up a forest estate of over 1 000 000 acres (405 000 ha). The area actually under plantations was 434 000 acres (175 630 ha) of which some 70 000 acres (28 327 ha) were acquired plantations; 714 000 acres (289 000 ha) were classed as plantable [**2**]. Private owners, with grant aid of £2 per acre for conifers and £4 per acre for broadleaves, had planted 126 000 acres (50 000 ha) [**3**].

Not all had been plain sailing, for the UK suffered the full force of the world-wide economic recession of the 1920s and 1930s. In March 1921 the Treasury stopped the Forestry Commissioners from acquiring more land following the recommendation of the Committee on National Expenditure (The Geddes Committee) that the afforestation policy adopted by Parliament after the war be scrapped. The Forestry Commissioners fought back and re-iterated the pro-afforestation arguments of the Acland Committee [**1**].

The four fundamental arguments on which forestry policy was based were, they claimed:

- The bulk of the reserves of mature coniferous timber (almost all in private hands) had been felled: most owners were unable to finance replanting.
- The world's virgin forests were being depleted.
- For national defence a reserve of standing timber within the UK was essential.
- Four million acres of land in the UK of little use for anything else was capable of growing timber.

Although they had to accept some stringent economies with cutbacks in the areas planted, in staff and in schools, the role of forestry as a rural employer suddenly came to the political forefront and the Commission was ordered to expand the afforestation programme that had just been cut.

The Forestry Commissioners of those days were not afraid of voicing their criticism of Ministers and Government.

They said in 1921 [**1**]:

> The reasons which rendered it necessary to have a Forestry Policy at all are not evanescent, but arise from a world-wide movement: the root causes of unemployment are not likely to disappear in a single season. On the other hand, successful forestry demands steadiness in policy, and it was one of the principal objects of the Forestry Act of 1919 to secure that continuity.

The performance of the private sector was regarded as "quite inadequate for maintaining private woodlands in a productive state". Some 450 000 acres (182 105 ha) had been felled during the war and afterwards [**2**] and the Acland Report had based its expectations on private owners re-stocking their woodlands with the aid of the financial assistance made available to them. High taxation of estate owners had altered their overall economic position and low prices for timber prevailing in the post-war years had acted as a deterrent to their actively rehabilitating their woods.

Numbers of estates were bought by "developers" (or "speculators" depending on one's point of view) and broken up and the assets sold off piecemeal; and standing timber was often sold to timber merchants for clear felling—no felling licences to clear woods were needed at this time and no replanting was required by law. The Commissioners in 1943 [**2**] described the treatment of the 450 000 acres (182 105 ha) in these words:

> "A small proportion was systematically replanted by the owners, a small part was acquired by the Commission for replanting, a considerable area was allowed to grow up to such species as came in naturally and much was left unplanted, subjected to grazing and is still more or less bare of woody growth. Meanwhile, felling continued without adequate replacement of the areas newly felled."

At one stage, the Commissioners thought that "in the long run the state might have to relieve the private owner of his responsibility in maintaining the woodlands of Great Britain and to take such work in hand before the woodland conditions have been dissipated, thus avoiding irreparable devastation".

One of the more interesting but less well-known recommendations of the Acland report [**4**] was that forest workers holdings should be established so that as in parts of Europe, notably Bavaria, the small holder could divide his time between farming (mainly a summer activity) and forestry (mainly a winter activity). The Forestry Commissioners were in agreement with this policy and decided that as a target, a forest worker's holding of up to 10 acres (4.0 ha) should be established for every 200 acres (80 ha) of plantations and that the holder should be guaranteed a minimum of 150 days work a year in the forest [**5**].

In the 1920s and 1930s forestry was, like farming, labour intensive from the time a crop was planted right through to the final felling. Indeed some operations that in farming were carried out by horse-drawn machines and later farm tractors could only be done by hand in the rougher often steeper forest terrain. Drains were mainly cut by hand as the crawler-tractor drawn forest plough had not been developed; planting on peaty soils involved cutting turves by hand and placing them (inverted) at the planting spacing; weeding of competing vegetation was, for the most part, by hand tools or hand-guided two wheeled machines.

The ubiquitous power saw of today was in use in primitive form only in North America and then only on a small scale; from Vancouver to Vladivostok timber felling in North America, Scandinavia, Europe and Russia was by man with axe and hand saw; timber extraction from the forest was by draught animal (mostly horses) or less commonly by powered winches and cableways; long distance movement of round timber was often by floating logs or rafts of logs down rivers. Conditions were tough indeed for the forest worker, especially when felling and extracting timber.

These were then mainly winter activities because snow could be compacted to winter roads and skid-trails became frozen especially over the boggy ground that was almost impassable in summer and that was so common in the Northern coniferous forests which stretch through Canada, Scandinavia and Russia and contain much of the worlds reserves of coniferous timber.

The Commissioners were keen to encourage public access to their forests. In 1927 steps were taken to make by-laws and to regulate public access under the Forestry Act 1927 [**6**] (primarily in the context of the management of the Ancient Royal Forests of the Dean and the New) and a committee was set up to investigate the feasibility of establishing Forest Parks. On its recommendation in 1935, the first National Forest Park was designated in Argyll [**7**]

In the Commissioners words "by taking a little additional expenditure in the utilisation of the land acquired for the new forests, it may be possible to provide, for the future, areas as highly prized by the public as the New Forest today." Similar thoughts were being expressed elsewhere in Europe.

In 1937 the Royal Scottish Forestry Society annual excursion took place in Bad Berka near Weimar in Thuringia [**8**], with the object of studying the silvicultural methods used in a large forest area managed under a general working plan made in 1728, which in 1828 was made into a very detailed plan, similar to the plan in force in 1937, with compartments and rides almost unchanged over the century.

In his speech welcoming (106 excursionists of) the Society, the senior representative of the Thuringian Forest Service is reported as saying:

> "In the new Germany today it is felt that, in spite of intensive forestry work, forests should not be regarded only as "Timber Factories" or as a source of revenue. They (the German foresters) wish to break away from the old artificial methods and by forming mixed woods to obtain natural woods which would not only yield good timber, but which, as a place of quietness and beauty, would satisfy the need of the people today."

It is later reported that the District Forest Manager (Herr Forstmeister de Marees):

> though desirous of keeping the number of roe deer low in order to avoid excessive damage, did not appear anxious to exterminate them entirely. He feels that a forest without wild animals would be unnatural and without one source of universal interest.

In Great Britain there was public concern about the preponderance of conifers in the Commission's planting programme and especially exotic conifers like Douglas fir and Sitka spruce whose rates of growth far exceeded those of the native Scots pine. But with its tiny reserves of timber and the generally poor economic situation, the country could not afford to do other than maximise the production of wood as the main aim of forestry. So the Commissioners were held on a tight rein by the Treasury, even though they would often have preferred to sacrifice some production of wood in favour of creating the more "natural" woods offering "quietness and beauty" and "satisfying the need of the people today", to quote the Thuringian Forest Service.

Of course, some concessions were made such as planting belts of hardwoods along roadsides in flat country and in strategic areas in hilly country. Liaison was established with the Councils for the Protection of Rural England and Rural Wales, and the Commission voluntarily declared some 'no go' areas for conifer planting e.g. in the core of the Lake District.

2 Inter-Departmental Committee on Utilisation (1931) and National Home Grown Timber Council (1936)

In 1931 the Forestry Commission appointed an inter-departmental home-grown timber committee to investigate and submit proposals for improving the utilisation of home-grown timber. Their interim report (1933) and a National Timber Conference of representative associations of owners, merchants and interested institutions led to the setting up in 1936 of the National Home Grown Timber Council as a vehicle for propaganda, the collection of statistics and the undertaking of limited amounts of research not undertaken by the Forest Products Research Laboratory, Princes Risborough. The Council was constituted as a limited company [**9**]. Its activities ceased on the outbreak of war and it was wound up in 1949. The Commission itself undertook investigations into the demand for timber in:

- Box and Packing Crate Manufacture in Great Britain, 1934 [**10**].
- Coal Mining in England and Wales, 1935 [**11**].
- Wood Turning in Great Britain, 1936 [**12**].
- Shipbuilding and Docks and Harbours, 1938 [**13**].

N.B.: The investigations into box and packing crate manufacture and into wood turning (in Scotland) were repeated by the post-war utilisation development committee. Other similar investigations covered the use and manufacture of wood charcoal; wood flour; timber used in river and sea defences; softwoods used in house construction; conifer bark and the yield and uses of sweet chestnut and of hazel coppice. The economic feasibility of establishing small-scale wood-processing plants was investigated [**14**].

3 Census of Woodlands 1938–1939

The virtual inevitability of war had by 1937 further concentrated the Government's mind on again having to greatly increase home timber production if war did in fact break out. The potential of forests to produce goods and services other than wood were given low priority as the Commission in collaboration with the Board of Trade was engaged in planning timber supplies in the event of war.

Private woodlands were surveyed and classified by a specially recruited team of young forestry graduates of whom at that time there was a substantial pool who were unable to find work at professional level [**15**]. This Census of Woodlands team, twenty or so strong, surveyed the private woods of England and Wales with their preponderance of broadleaves in late autumn and the winter when the leaf is off the tree and visibility is at its best and moved north to Scotland and northern England's mainly coniferous woods in other seasons. The 1919 Forestry Act gave the Commission powers to enter upon private land. As the team moved from county to county news of their arrival was given wide publicity and the number of encounters with irate landowners, farmers and householders was small and the greatest danger to the surveyors with their maps and instruments lay in their being reported as spies[1].

4 Summary Inter-War Years

The main achievements of the inter war years were:

- The Forestry Commission planted 369 000 acres (150 000 ha) Some 90% of the species planted were conifers.
- Private woodland owners with grant aid planted some 126 000 acres (50 000 ha). Little other planting was done.
- 1471 forest workers holdings were established.
- Forester Training Schools were established in the Forest of Dean (where there had been a training school for woodmen prior to 1914); at Beaufort, Inverness, on Lord Lovat's estate from 1922 until it was transferred in 1930 to Benmore Argyllshire; small schools were opened for short periods in the New Forest (1922) and at Chopwell, Co. Durham, in the 1920s.
- A substantial programme of research was set in train, mainly in relation to silviculture.
- Four British Empire Forestry Conferences had been organised (GB 1920; Canada 1923; Australia and New Zealand 1928; and South Africa 1935).
- Three National Forest Parks were opened—Argyll in 1935, Snowdonia in 1937 and Forest of Dean in 1939.
- Early in 1939 the Commission set up the Home Grown Timber Advisory Committee representing the home timber trade, private woodland owners, the Board of Trade and the Commission. With war imminent, one of its first tasks was to set maximum prices for standing timber, round timber in log form, sawn timber and mining timber; and to draft orders to control the felling of standing timber. These regulations came into force on the outbreak of war in September 1939.
- And last but not least, a country-wide administration was in place to deal with all the many facets of an expanding planting programme from raising plants through land acquisition to plantation establishment; the design and implementation of a private woodlands grant aid scheme; the organisation of the Empire Forestry Conferences; and the Forest Authority role as Government advisor on forestry policy for both State and Private sectors and Department of Timber Control in the event of war.

[1]Author's Footnote: In July 1939 an embarrassed local police sergeant had to call on my parents home in Inverbervie, Angus, to apprehend me as a possible spy, although he was well aware that I was by then serving in the Royal Engineers in Catterick having been called up with the first (and only) batch to do 6 months peace-time National Service under the Militia Act of 1939: when war broke out in September the Act was replaced by full emergency powers and the Militia was incorporated into the war-time army.

When war was declared in September 1939 the Commission was divided into two Departments [**1**]:

- The Forest Management Department to carry the normal functions of the Commission.
- The Timber Supply Department to deal with war demands for timber.

In February 1941 the Commission handed over its Timber Supply functions to the Timber Production Department of the Ministry of Supply.

References

1. *The Forestry Commission. The First 75 Years.* 1994
2. *Forestry Commission Post War Forest Policy Report by H.M. Forestry Commissioners.* Cmd 6447, June 1943. Reprinted 1956
3. *30th CAR*
4. Ministry of Reconstruction. Reconstruction Committee Forestry Sub-Committee *Final Report (Cd. 8881).* HMSO 1918
5. *4th CAR* (1924)
6. *Forestry Act 1927*
7. *16th CAR* 1935 Argyll Nat. For. Park
8. *The Scottish Forestry Journal Vol 51, Part 2.* Report on the 55th Annual Excursion (1937) A.H. Gosling
9. Forestry Commission. *Report of the Committee on Marketing of Woodland Produce 1956.* HMSO
10. Forestry Commission Utilization Series No. 1. *Report on the Demand for Timber for Box and Packing-case manufacture in Great Britain, 1934.* HMSO
11. Forestry Commission Utilization Series No. 2. *Report on the Demand for Timber in Coal-mining in England and Wales, 1935.* HMSO
12. Forestry Commission Utilization Series No. 3. *Report on the Demand for Timber in Wood-turning in Great Britain 1936.* HMSO
13. Forestry Commission Utilization Series No. 4. *Report on the Demand for Timber in Ship building and in Docks and Harbours 1938.* HMSO
14. Forestry Commission. *Forest Record 19 The Manufacture of Wood Charcoal in Great Britian.* HMSO 1953
    Forestry Commission. *Forest Record 21. Use and Manufacture of Wood Flour* HMSO 1953
    Forestry Commission. Forest Record 27. *Use of Home-grown Timber in Wood Turning and Related Trades in Scotland in 1953.* HMSO 1954
    Forestry Commission. *Forest Record 29. Use of Forest Produce in Sea and River Defence.* HMSO 1954
    Forestry Commission. *Forest Record 30. Growth and Yield of Sweet Chestnut Coppice.* HMSO 1956
    Forestry Commission. *Bulletin No. 27. Utilisation of Hazel Coppice.* HMSO 1957
    Forestry Commission. *Forest Record 35. The Use of Home-Grown Timber in Packaging and Materials Handling.* HMSO 1958
    Forestry Commission. *Small Pulp Mill Survey: Economic Study.* HMSO 1959
    Forestry Commission. *Board Mill Survey. Economic Study.* HMSO 1960
15. Forestry Commission. *Census Report No. 1 Census of Woodlands 1947–1949*, p. 15. HMSO 1952

Chapter 2

1940s Post (1939–1945) War Policy

1 The June 1943 and January 1944 White Papers

Two-thirds of the timber standing in British woodlands in 1939 had been felled by 1946 to meet war-time needs. The blockade at sea had virtually cut off supplies from a number of wood-rich countries that traditionally supplied timber in peace time—such as those bordering the Baltic Sea—and shipping could ill be spared to bring wood and wood products from other more distant countries.

As early as 1943 the Forestry Commissioners had a clear idea as to what should be their post-war forest policy and their proposals were presented in a report "Post-War Forest Policy" in February 1943 [**1**] to the then Chancellor of the Exchequer, who presented the report to Parliament in June 1943 as a White Paper (Cmd 6447).

The White Paper proposed that 5 million acres (2 million ha) be devoted to forests systematically managed and developed. That area, said the White Paper, "is required for national safety and will also provide a reasonable insurance against future stringency in world supplies. There are valuable contingent advantages associated with forests, such as the development and settlement of rural Britain."

Although it is evident from subsequent statements in Parliament that politicians were greatly influenced by the need to build up a strategic reserve of growing timber—the 5 million acres was linked with the needs of a 3-year war—the Commission proposals to Parliament quite specifically drew attention to amenity and recreation.

The National Forest Parks set up in the 1930s by the Commission were considered to be a success and the Commissioners considered that several other Commission owned areas were eminently suitable for parks. The opening up of forests generally to youth organisations had begun and much more was envisaged including the provision of basic facilities such as water and sanitation at semi-permanent camp sites.

Although the provision of these recreational facilities by the Commission attracted support from all the Governments of the day, it was the post-war Labour Government that gave most support to the idea. In 1946 Hugh Dalton (Chancellor of the Exchequer) on a visit to North Scotland Conservancy described his vision of coach loads of factory workers visiting Glen Urquhart and Glen Affric Commission forests to clear their lungs of the polluted air of the Black Country. The beauty and tranquillity of the remote countryside, hitherto the perquisites of the better off, would become available to all. As it turned out, in due time the Commission's forests would come to provide millions of people with an opportunity to take fresh air and exercise, albeit the bulk of those enjoying the forest would be arriving by car. The need to reconcile the claims of amenity and economic utilisation were regarded as "almost a truism in these small islands."

It was estimated that 2 million acres (800 000 ha) could be secured from the existing woodland area (much of which had been felled within the previous 30 years or so) by selecting those woodlands better suited for forestry than for any other national purpose. The other 3 million acres, to be secured by planting of bare land, mainly by the State, would be afforested gradually; the loss to food production was thought to be small since only poor agricultural land would be taken.

To secure systematic long-term management of the 2 million acres of existing woodlands, most of which were in private ownership, a Dedication Scheme was proposed. Dedicated woodland would be worked to a plan of operations approved by the State and in return owners would receive up to 25% of their net expenditure on those woodlands, up to the time the woodlands were self-supporting. Loans at current rates of interest would also be available.

This proposal, which became known as Basis I Dedication, was not popular because of the complicated very detailed accounting methods of arriving at the net expenditure on woodlands in what for the most part were mixed agricultural and forestry estates and subsequently an alternative scheme (Basis II) was to become the main vehicle for implementing Dedication.

For the remaining substantial areas of small woods, shelter belts and amenity plantings, free technical advice would be provided but no State grant aid. The Government said that there could be no decision in respect of the Private woodlands proposals made by the Commission until the views of private owners had been further explored. Accordingly, the Commissioners held formal meetings with the private Associations and Societies interested in forestry and in November 1943 presented a Supplementary Report on Private Woodlands [**2**] to the Chancellor of the Exchequer; this was presented to Parliament in January 1944 (Cmd 6500 HMSO January 1944).

The Supplementary Report made it clear that (the conference of) private owners were concerned that the Forest Authority might move too quickly for some owners, either in demanding immediate decisions on Dedication or in forcing a pace of work in the woods for which owners would not be geared up immediately after the War.

2 Private Woodlands: The Dedication Scheme Agreed

After much debate, the alternative basis of a simple planting and maintenance grant (Dedication Basis II) was unanimously agreed. The Conference accepted the assurances given by the Commission that, given goodwill on both sides, it should be possible to avoid the difficulty of the Commission forcing the pace of Dedication and that of the indifferent owner using delaying tactics.

The Conference agreed on seven fundamental principles (Cmd 6500, page 4 paragraph 6) [**2**]:

1. That the rehabilitation of the woodlands must proceed with both certainty and rapidity.
2. That all planting and natural regeneration acceptable instead of replanting must be properly looked after up to the stage of satisfactory "establishment".
3. In every case the owner must reach an early decision as to whether he is prepared to proceed with the work of rehabilitation.
4. If an owner is so prepared and can give satisfactory assurances he is deserving of financial assistance from the State.
5. Where no satisfactory assurances are forthcoming within a stated period, the State should acquire the land.
6. That the war-time system of felling licenses must continue until the reserve of standing timber can be adjudged satisfactory.
7. That some degree of control of the silviculture of private woodlands is necessary.

It was a considerable achievement for the Commission to secure the Conference's agreement to its proposal that all woodlands judged to be suitable and necessary for timber production should either be Dedicated to that purpose by the owner or be acquired by the State. Dedication would run with the land and would be unaffected by changes of ownership.

The owners representatives also secured undertakings and clarifications of the Commission's intentions. In particular, small woods suitable for timber supply purposes but not suitable for Dedication would get a planting grant, though not a maintenance grant. Small woods could, however, be drawn into the Dedication scheme if they were in company with larger woods and so enjoy the Dedication scheme grants.

The role of amenity woods—and especially small woods—was recognised, but the Commission was adamant that it was not the province of the Forest Authority to spend on amenity plantings per se money placed at its disposal for forestry purposes by Parliament. However, joint action with the Planning Authorities was envisaged and in the case of shelter belts, with the Agricultural Departments.

The two Forestry Societies (the Royal Scottish and the Royal English) proposed that there should in effect be separate organisations to administer State and Private Forestry, but this and other proposals were resisted by the Commission on various grounds, not least that to attract the best brains the Forestry Commission would require to be of substantial size with good career prospects if it were to attract top professional foresters. The Commission felt that the risks of Commission domination of private interests could be reduced by the appointment of local Advisory Committees on which woodland owners and others would be represented. The notion of a large central consultative committee was not favoured.

The introduction of the Dedication Scheme (Basis I and Basis II) was approved by Government in November 1945. It was 1948 [**3**] before all the legal and other details were finalised and the scheme became fully operational.

The discussions and the delays were fully justified, for once accepted by the private sector it became the vehicle by which the private sector invested in its massive post-war planting programme.

3 Changes in the Forest Resource 1919–1947: Progress of Planting New Woodlands and Re-stocking Felled Areas

It may be useful here to record the progress of replanting felled woodlands and planting new woodlands in the period 1919 to the end of the second world war.

3.1 Private Woodlands

The losses of woodland due to the first world war fellings have already been recorded. Although, with the aid of planting grants and unemployment relief grants, some 126 000 acres (50 000 ha) had been planted by private owners in the two decades 1919–1939 [**1, 4**], most of the planting was the re-stocking of felled woodlands and after allowing for the area of woodlands felled in this period and not replanted, there was an overall loss of woodlands.

In their 30th CAR the Forestry Commission noted that "although the statistics are not very reliable, it was estimated that to meet war-time needs about one-seventh of the mature woodlands had been laid low" [**4**].

From 1940 to 1946 inclusive some 27 700 acres (11 210 ha) were planted [**4**]. Proposals for the Dedication Scheme—the main grant-aid scheme for the rehabilitation of private woodlands was approved in 1945 but progress in implementing the scheme was slow due to time needed to agree the legal aspects of the scheme with private owners representatives and to incorporate agreed alterations in the Dedication deeds—such as a clause to allow owners to leave the scheme in exceptional circumstances.

By the end of September 1949, the scheme began to take off. Owing to delays involved in formulating the legal details of the scheme, from 1946 owners could apply for an Interim planting grant.

The actual planting undertaken in the three years 1947–1949 is estimated to have been (Table 11 in Ref. [**4**]):

Table 2.1 Progress of Private Planting 1947–1949 (acres)

Forest Year	Total	Grant Aided	Other
1947	9000	7600	1400
1948	11 100	8800	2300
1949	13 300	10 200	3100
	33 400	26 600	6800
(ha)	(13 516)	(10 764)	(2752)

3.2 State Forests

The areas planted by the Forestry Commission from 1920–1947 (years ending 30th September) were (Table 33 in Ref. [**4**]):

Table 2.2 Areas planted by the Forestry Commission (acres)

Year	Planted Afforested	Re-afforested	Total Acres	(Hectares)
1920–1929	101 976	36 295	138 271	(55 955)
1930–1939	174 428	56 179	230 607	(93 322)
1940–1947	95 072	41 760	136 832	(55 375)
Total	371 476	134 234	505 710	(204 650)

The total area planted 505 710 acres (204 650 ha) was divided between conifers and broadleaves as follows:

Table 2.3

	Conifers	Broadleaves	Total Acres
Afforested	357 175	14 301	371 476
Re-afforested	110 400	23 834	134 234
	467 575	38 135	505 710
(ha)	(189 218)	(15 432)	(204 650)

In CARs post-1947, the term afforested may be replaced by new planting and re-afforested by re-planted or re-stocked. Afforestation, the planting of land not previously carrying a crop of trees and re-afforestation (= re-forestation), the planting of a former woodland site, are FAO's European Forestry Commission terms agreed on in early post-war years for international use.

Table 2.4 Total area of land acquired through the Forestry Fund as at 30th September 1947 (acres) [**5**]

Forest Land	England	Scotland	Wales	Great Britain
Acquired Plantations	40 100	11 900	3200	55 200
Planted by FC	211 100	182 400	85 400	478 900
Await planting	109 800	191 400	45 200	346 400
TOTAL FOREST LAND	361 000	385 700	133 800	880 500
Nurseries	900	700	200	1800
Agricultural	17 400	32 400	12 800	62 600
Forest Workers Holdings	6700	2900	4200	13 800
Unplantable and miscellaneous	102 500	349 400	29 900	481 800
GRAND TOTAL	488 500	771 100	180 900	1 440 500
(ha)	(197 685)	(312 047)	(73 206)	(582 938)

The 1947–1949 Census of Woodlands (see below) gives the area of Forestry Commission Woodlands (State Forests) as 534 491 acres of High Forest as at 30 September 1947, which is in close agreement with the 534 100 acres of "acquired plantations" plus "planted by FC" in the table above

For a variety of reasons, the acquisition of land for planting inevitably involves the purchase of other land not suitable for planting which may be interspersed with plantable land. The Commission has nevertheless had to justify to the forestry Ministers and to Parliament its holding of such a high proportion of non-forest land as, for example, is shown in the table above, viz., 62 600 acres agricultural land and 481 800 acres of unplantable land (of which 1800 acres were forest nurseries and 13 800 acres were forest workers holdings) as compared to 880 500 acres of forest land.

4 The Census of Woodlands 1947–1949

The Census of Woodlands of 1947–1949 [**6**] was the first ever comprehensive survey of woodlands in Great Britain. It included surveys by the Commission of its own land and a survey of all private woods over 5 acres in extent by the specialist Census of Woodlands team recruited by the Commission.

Table 2.5 Area of High Forest in Great Britain by age classes, composition and ownership (acres × 1000)

At 30th September 1947				Woods of 5 acres and over		
Age class	Mainly broadleaved			Mainly conifers		
	Private	State	Total	Private	State	Total
Total	777.3	65.5	843.8	476.9	468.0	944.9
Over 80	195.2	24.4	219.6	37.1	3.6	40.7
61–80	81.3	2.1	83.4	45.5	3.1	48.6
41–60	49.0	1.9	50.9	61.5	3.8	65.3
21–40	49.2	4.9	54.1	143.1	79.7	222.8
0–20	18.8	23.8	42.6	134.2	365.6	499.8
Uneven aged	383.8	9.4	393.2	55.5	12.2	67.7

Table 2.5 shows that in 1947 in broadleaved woodlands (High Forest) in which the older age classes predominated the State owned less than one-twelfth of the total area. In conifers (High Forest) which consisted largely of the younger age classes, the State owned almost half the total area [**7**].

High Forest = stands of trees of any age forming a forest canopy or capable of doing so—as for example newly planted woodlands.

5 Private Woodlands—Taxation

Historically, the taxation of private woodlands has recognised the long time span between an owner incurring the expense of establishing and maintaining a woodland and (very often) his heirs enjoying the income from the sale of the mature timber anything between 50 and 250 years later.

No attempt is made here to describe the technical details of those forestry tax regimes. One effect was to enable seriously wealthy individuals to offset their expenditure on establishing new woodlands against their income from other sources and to avoid or reduce drastically any income tax payable on receipts from timber sales as the woodlands became revenue earning and duty payable on the woodland estate at death. But it was not until the 1950s that a number of forest management companies set out to help woodland owners and would-be owners to take best advantage of these favourable forestry tax regimes. Coupled with the grants payable under Dedication and Approved woodlands schemes, the effect was to increase private planting.

Private planting—including some restocking of felled woodlands—rose from just over 40 000 acres (16 187 hectares) in the five year period 1945–1949 [**4, 5**] to some 83 000 acres (33 588 hectares) in the five years 1950–1954 and 147 700 acres (59 771 hectares) in the 5 years 1955–1959 (see 31st to 40th CAR). The scale of some of the new plantations and the conspicuous wealth of some individuals creating them caused considerable public criticism of what many saw as 'tax avoidance' schemes. In reality some of the poorest land in Britain was being turned into highly productive forests and forestry remained the poor cousin compared to agriculture in respect of government subsidies.

After various changes consequent on general changes in taxation (e.g. capital gains tax, inheritance tax, business tax relief) in 1988, and without consultation, the then Prime Minister ordered that forestry be taken out of the tax system altogether. The effect on private woodlands planting is discussed later (see also Ref. [**8**]).

6 Staffing of the Post War Forestry Commission

On the outbreak of war in 1939 the Commission took charge of home timber supply [**9**]. In February 1941 it handed over this responsibility to the Timber Production Department (TPD) of the Ministry of Supply. Commission staff were seconded to the TPD for the duration of the war; a small staff continued with a limited programme of new planting and restocking and the essential maintenance of the Commission's estate—woodland and other land. At the end of the war, the large post-war planting programme of the Commission and the implementation of the grant-aid scheme (Dedication) required a large staff to be recruited. Fortunately forestry offered a career which was attractive to many of the men being demobilised after war-time service in the armed forces. Some who had embarked on other careers before joining up, on demobilisation took advantage of the forestry training and educational opportunities offered to them at Universities and the Commission's own forester training schools. There was no shortage of jobs for forest workers—often with the prospect of a Commission house, sometimes with an agricultural small-holding which carried with it a guaranteed number of days work in the forest.

Forestry benefited from the large intake of managers, supervisors and forest workers with experience of service life and in some instances of quite other pre-war occupations.

Together with the return of staff seconded to the TPD and the nucleus of Commission staff who had remained in the Commission during the war they constituted a body well suited to the challenging task of organising the establishment of the very large areas of plantations which the post-war governments forestry programmes demanded.

7 The Marketing of Forest Produce

In 1943, two potentially conflicting principles were discussed: the building up of a reserve of standing timber and the maintenance of an efficient home timber trade as the nucleus for expansion in times of emergency. The compromise suggested was that whilst the war-time control of felling of trees be continued after the war with a reduction in the amounts to be licensed for felling, the Commissioners would have increasing quantities of thinnings to sell and would continue to dispose of their timber in the round to the trade except where no satisfactory offer was forthcoming or where the Commission wanted to carry out pioneer work such as the promotion of some new industry. The consequence would still be a reduction in the size of the war-time home timber trade, but with the prospects of re-growth and further expansion as the large new forestry planting programmes began to yield increasing quantities of thinnings.

One of the key phrases in the forest produce section of the White Paper [**1**] was "It will be an important function of the Forest Authority to foster new industries"; this was seen by some of the established home trade as providing the Commission with an excuse to by-pass them as round timber merchants and to engage directly in the felling and transport of timber to new industries set up to process small round wood from thinnings. But in fact there was room enough and need enough for both the Commission and the timber trade to harvest wood to meet the needs of the new forest products sector.

8 Forest Research

The 1919 Forestry Act empowered the Commissioners to "make or aid in making such enquiries, experiments and research and collect or aid in collecting such information as they may think important for the purpose of promoting forestry and the teaching of forestry and to publish or otherwise take steps to make known the results of such inquiries, experiments or research and disseminate such information."

To advise it in planning and carrying out forest research the Commissioners appointed, in 1930, an Advisory Committee on Forest Research [**4**].

Due primarily to lack of funds, the Commission had no central research station. The research officers, who were stationed in various parts of the country, carried out investigations involving field studies. The more elaborate investigations requiring specialised techniques and laboratory equipment were farmed out to Universities or

Research Institutions. This policy worked well and it was suggested that it should continue. It was recognised, however, that it was proving more and more difficult for the Commission research officers to provide an efficient service to management solely from field studies and without the backing of the facilities of a Forest Research Station. Accordingly, it was proposed to implement a 1920 Cabinet Sub-Committee [**10**] recommendation that the Commission establish a forest research institute. It was only in 1946 that this was done (at Alice Holt Lodge, Surrey) but in the meanwhile the Commission continued to organise forest research for its own needs and for the needs of the private sector and continued to conduct much of that research with its own staff. In this Britain was very much the exception, for in many countries forest research was divorced from the forest service and it is often said, from reality.

9 Forest Products Research

In 1920, the Agricultural Sub-Committee of the Cabinet Committee had assigned research into the utilisation of timber and other forest products to a Forest Products Research Board under the Department of Scientific and Industrial Research (DSIR). Its Forest Products Research Laboratory at Princes Risborough—which also dealt with timbers from the then colonies—became in the course of time one of the most prestigious of the world's forest products laboratories (in the 1970s it was moved and became a part of the Building Research Establishment).

10 The 1945 Forestry Act—National Committees—Regional Advisory Committees—Forestry Commission Re-organised

Under the Forestry Act of 1919, executive control of the Forestry Commission had been entrusted to three Assistant Commissioners—one for England and Wales, one for Scotland and one for Ireland. Forestry in Ireland passed to the Provisional Government of the Irish Free State and the Government of Northern Ireland in April 1922 [**11**]. The Assistant Commissioner for England and Wales and the Assistant Commissioner for Scotland operated through Divisional Offices—seven in England and Wales, including the Deputy Surveyorships of the Dean and the New Forest and three in Scotland.

One of the main purposes of the 1945 Act [**11**] was to give effect to changes in the organisation and powers of the Forestry Commission which they had been given in 1919. In particular, the Act placed responsibilities which had hitherto been exercised by Commissioners on Forestry Ministers i.e. the Ministers of Agriculture for England and Wales and the Secretary of State for Scotland (when Wales got its own Secretary of State he became the Forestry Minister for Wales). The Commissioners, in exercising their functions under the Forestry Acts, had to comply with any directions given then by the Forestry Ministers. Thus the Commission became in effect a Government Department, not an independent Commission.

National Committees were set up for England, Scotland and Wales respectively to carry out duties devolved on them by the Commissioners relating to land acquisition, cultural operations, estate management, National Forest Parks and private forestry.

The Committee membership comprised three or four full-time Commissioners and up to three non-commission members. The Forestry Commissioners were empowered to devolve duties to National Committees.

The National Committees held their first meetings in November 1945. In the 28th CAR 1947 [**5**] it is recorded that in that year the three National Committees met monthly except August and had under review land acquisition, the planting programme, housing, private forestry and forest workers training.

Under the Act, Members of Parliament were no longer eligible as Commissioners. The National Committees were supported and given advice at Conservancy (see below) level by eleven Regional Advisory Committees which first met in May or June 1946.

To assist the Commissioners in the execution of their policies and for administrative and technical co-ordination, the posts of Director-General and Deputy Director-General were created together with a Controller of Finance, a Director of Research, Education and Publications and an Information officer.

Directors of Forestry were appointed for England, Scotland and Wales. They in turn operated through regional Conservators (five in England, two in Wales and four in Scotland) and the Deputy Surveyors of the Forest of Dean and the New Forest. Each Conservator and Deputy Surveyor was an accounting officer with his own annual budget and responsible for compiling and submitting to the Director General an annual financial statement of his stewardship.

Conservators were assisted by two senior professional forest officers of Assistant Conservator grade, whose responsibilities were divided on a territorial basis; by a Chief Clerk with a substantial team of administrative staff; and by specialists such as a Land Agent, a Civil Engineer and a Mechanical Engineer.

Conservancies were divided into Districts each under a District Forest Officer. A District comprised several Forests, each managed by a Forester-in-charge.

The Conservator was responsible for giving advice in the field to the private woodland owners in receipt of grant aid and for the administration of the grant aid scheme. This could and did give rise to problems in setting priorities between on the one hand dealing with the Commission's own programme of work and on the other duties to the private sector. This problem was eased by the appointment in 1947 of Private Woodland Officers in each Conservancy, but with tight control of staff numbers, there remained the need for field officers to devote part of their time to private woodlands advisory work. Many landowners welcomed the contacts with Commission foresters and District Officers with day to day experience of forest management in their locality.

This type of multi-layered management structure was very much in vogue in the post-war years—for example in the National Coal Board. So far as the Commission was concerned it had the great advantage of having local staff in charge of the actual field operations with detailed and in-depth knowledge of local conditions, particularly those affecting the establishment of woodlands. This was of course of major importance when one considers the wide range of soils and micro-climates on which the Commission were yearly establishing many thousands of hectares of new woodlands in the post-war years, and advising the private sector on its increasing afforestation programme.

The main disadvantage was the long chain of command—an instruction given by the Commissioners passed through 3 or 4 levels of command (for example, the Director for the country, the Conservator, the Assistant Conservator, the District Officer) before being implemented by the Forester.

Apart from being time-consuming (and communication was generally by manually typed or hand-written letter or 'memorandum' in those days) there was the problem of how far to allow the different levels of management—Country, Conservancy, District—to adapt the original intent of the Commissioners to local circumstances without distorting the Commissioners original intent.

One of the first public bodies to address this type of management problem was the National Coal Board and their decision to cut out one complete level of management was much discussed within Commission Headquarters as a possible role model for it to emulate. No major change was made until 1965 when the Directorates for England, Scotland and Wales were abolished and Headquarters dealt direct with Conservancies.

The main work of the Commission from 1919 up until the 1950s was almost entirely concerned with acquiring and afforesting bare land and the management of the Dean and the New Forest. But a new major task was on the horizon—that of harvesting and marketing the rapidly growing quantity of conifer thinnings from the earlier plantings and helping private woodland owners with marketing their conifer thinnings—now also increasing in quantity.

Secondly, as we have already noted, the administration of the private woodland grant-aid schemes—mainly the Dedication and Approved Woodlands Schemes—was putting more and more pressure on the Commission staff at all levels.

It may be pertinent here to look ahead to the 1960s when serious thought was given to making the administration of private woodland grant aid the responsibility of a separate "Forestry Authority" with its own offices and administrative set up and a budget given to it by the Forestry Commissioners, but the idea never took off. Private Woodland owners, a much more powerful lobby with much more voting power than in the 1980s and 1990s were opposed to it because they felt that they got better advice from staff of a body practising forestry in the local (Commission) forests than they would from a purely advisory service whose staff would in time lose touch with the best hands on forestry and the best forestry practices—and it has to be remembered that foresters in both State and Private sectors were on a steep learning curve as new methods, machines, chemicals were being developed by the research branch of the Commission, by its work study (operational efficiency) branch, by machinery manufacturers and field practitioners themselves.

Equally it was felt by many that forestry was too small a profession in Britain to split up into two separate bodies, each with its own career prospects. As has already been mentioned, the Commission in the immediate post-war years had attracted a lot of very talented people, a good proportion being ex-servicemen who had other careers before their war-time service or had had distinguished war-time records. Better to keep them together the argument went than to split them into two services—State Forests and Private Woodlands.

In time it became clear that within a unified practising and advisory service there was room for and indeed, a real need, for specialisation with a role for general silviculturists/forest managers and one for specialists in harvesting and marketing. Changes did in fact come about, albeit gradually.

11 Government Policy Statements of 1945 and 1946

Forest policy was the subject of an announcement in the House of Commons on 30th November 1945 [**5**] and again on 24th October 1946 [**5**] when the House debated the motion: "That this House approves the policy of His Majesty's Government for Forestry as announced by the Minister of Agriculture and Fisheries on 30th November 1945".

In his opening statement, the Minister gave approval to the Commissioners proposals; "after careful and detailed study the Government were impressed with the necessity, as a safety measure, apart from all other considerations, to rebuild our reserves of standing timber as quickly as possible." He went on to speak of the better utilisation of poor land, increased rural employment and the improved recreational facilities that would result from large scale action by the Forestry Commission. The Government, he said, also recognised the importance of private afforestation and accepted the Dedication scheme.

He paid tribute to the landowners of 50 to 70 years ago, through whose foresight and initiative in planting trees and creating and managing woodlands, there was timber enough to carry the country through two world wars.

12 The 1947 Forestry Act

In 1947 a Forestry Act received Royal Assent; its purpose was to get over certain administrative legal difficulties which had arisen in connection with the dedication of land for forestry purposes [**5**].

13 Forestry Educational Centre, New Forest

Northerwood House, a Georgian mansion on the outskirts of Lyndhurst (New Forest) was gifted to the Forestry Commission for use for forestry education—in the broadest sense of the phrase. The 1947 CAR [**5**] notes that by June of that year it had been adapted for use as a hostel for holding short courses of instruction for officers of the Department (Forestry Commission), for university students doing practical training and for conferences generally. Over the years it was also used for short forestry courses for school teachers, landowners and land agents and timber merchants.

The first occasion on which it was used was for a conference of the heads of delegations of the Fifth British Empire Forestry Conference which took place in 1947 [**5**]. It is doubtful if anyone associated with the early days of Northerwood House had any idea how successful it was to become as a centre "for promoting forestry" where large numbers of individuals would enjoy a week's intensive course on some aspect of forestry in the setting of the New Forest with its wide range of woods of different ages and species. The courses were conducted by practising Forestry Commission staff who could talk with "hands on" experience. The non-institutional country house setting gave an informal friendly background against which to study. Sadly, for a variety of reasons, the Commissioners decided not to undertake the extensive restoration of Northerwood House which became urgent by 1970 and the building was sold.

A training suite was incorporated in the Edinburgh office to which the Commissioners moved from London in 1975.

14 International Co-operation

As has been noted in the Introduction, in the 1919 Act the Commissioners were given powers "to make or aid in making such inquiries as they think necessary for the purpose of securing an adequate supply of timber in the UK, and promoting the production of timber in His Majesty's Dominions"; the First Empire Conference was held in Britain in 1920. The Fifth British Empire Forestry Conference was again held in Britain, in 1947, when 33 countries were represented. Some 66 technical papers were considered and the 112 Conference delegates toured Britain in a special sleeper train which acted as their mobile hotel.

Representatives of the Food and Agricultural Organisation of the United Nations attended, to explain the work of the FAO Forestry Division.

Britain played a key role in encouraging international co-operation on forest policy, forest management techniques and forest research. It was realised that with a large ambitious forestry programme, there was everything to be gained from discussions with countries with long forestry traditions and that the pooling of knowledge, experience and ideas could be of real benefit to Britain. Thus Forestry Commission

Officers played a leading role in establishing the European Forestry Commission of FAO, which, working with the Economic Commission for Europe (ECE), was to assist Governments to formulate and adopt co-ordinated forest policies in the aftermath of the Second World War.

There had been World Forestry Congresses in 1926 and in 1936. In 1949 a third Congress was held in Helsinki [**4**] when a Forestry Commissioner was elected Vice-President of the Congress. Britain continued to play an important role in succeeding World Forestry Conferences.

In 1947 a Forestry Commissioner attended a Conference in Czechoslovakia convened by the Food and Agriculture Organisation (FAO) of United Nations which led to the establishment of the European Commission for Forestry and Forest Products, now known at the European Forestry Commission (EFC) [**4**].

Britain also played a leading role in the post-war re-building of the International Union of Forest Research Organisations (IUFRO), a long established (1890) and effective world wide research "club" of research institutes; and in creating new advisory and specialist research organisations like the International Poplar Commission of FAO.

The Twelfth Congress of IUFRO met in Oxford in 1956. It was attended by 250 delegates from 42 countries [**12**]. The Director of Research and Education was appointed President. IUFRO has gone from strength to strength and is the major international research organisation on the world forestry scene.

15 Rural Employment and Housing

In the Forestry Commission, the Government of the day had an organisation that could implement rural employment and rural housing policies, whilst pursuing its goal of building up as rapidly as possible a national reserve of timber in State and private ownership. At various times employment schemes had been tried—some on private initiative.

Thus after the First World War, Lord Leverhume bought land on the Island of Lewis and tried to settle ex-servicemen on small holdings. But the idea and his management style was alien to the islanders' culture and customs. In contrast, the large acquisitions of plantable land throughout Britain by the Forestry Commission gave opportunities for regular paid employment which were eagerly taken up, in spite of the arduous nature of the work. In passing it is worth noting that key to the on-site training of these workers and responsible to the forester in charge of the forest area was the foreman, an experienced forest worker, who in turn would have under him one or more gangers each leading a small squad of men, whilst fully participating in the work.

At the end of the Second World War, Forestry Commission acquisitions of bare land for planting increased rapidly—much of this land was in the uplands, often of poor quality degraded by sheep grazing or used for sporting as in the so-called (red) deer forests of Scotland. Sometimes, whole estates including arable farm land and better quality sheep grazing land were acquired; the plantable land was managed separately from the other land which was let to agricultural tenants.

The numbers of forest workers required to undertake the planting programmes often far exceeded the availability of local labour. In addition to the building of several 'forest villages' on green field sites, many small housing developments were undertaken. The new permanent houses were an improvement on the pre-war agricultural cottages. Ex-servicemen, often newly married, seeking a home and work were eager tenants. Quite a proportion were from towns. Some settled far from their places of origin and English accents could be heard in the Highland glens of north Scotland.

In remote areas, the rate of building permanent houses in forest villages and in smaller communities (of, for example, half a dozen houses) was slow compared to the targets set (CAR 1949 [**4**] and the remarks of Hugh Dalton, then Chancellor of the Exchequer, in his memoirs [**13**]).

Recourse was made to building "temporary" houses. Many of these were made up of sections of wooden army, Canadian Forestry Corps, and air-force barracks salvaged by demolition contractors and re-erected to give small two and three bed-roomed houses. They were surprisingly comfortable and heat efficient due to their being given an inner lining of plaster-board and an outer cladding of pebble dash on wire mesh or, even better, a cladding of Canadian cedar shingles. And many that were eventually sold as holiday houses lived on for many a year longer than had been intended when they were built.

Another expedient whilst houses were being built was to house men in bothies. These were also built from ex-army huts re-located by specialist contractors and furnished like barrack rooms with ex-War Department equipment, but often with individual cubicles rather than the open-plan of the traditional barrack room.

A kitchen, dining room and sometimes a recreation room completed the picture. If there was no local bus service, once or twice a week the forest lorry would be fitted with wooden benches and would take workers and their families to the nearest village or small town for shopping. Pay day was weekly on a Friday. In remote forests on Wednesday, the forester in charge would make up the pay sheet and send it to his Conservancy office, which on Thursday would send the cash by registered post to the Post Office nearest to the forest for delivery or collection by the forester on Friday, in time for the four o'clock pay out.

It is amazing how well the system worked, considering the remoteness of some of the forests from Conservancy offices. For example, the North Scotland Conservancy office in Inverness served forests on the Islands of Mull and Skye and at Borgie on the Pentland Firth—none of which had a railway station and which therefore depended on the post bus and for the islands a ferry crossing for the last part of the journey.

Mostly forests were grouped into Districts of up to a dozen forests (the largest forests such as Kielder in Northumberland and Thetford in East Anglia might themselves comprise a District, being divided into sections under the supervision of a forester).

One of the duties imposed on District Officers by the Treasury via the Commission's Controller of Finance, was to carry out "surprise checks" on pay day and to witness the handing over of the pay and pay slips to each employee. When the "surprise check" involved the pre-booking of a local hotel (and maybe a ferry as well) as in the Highlands of Scotland during the salmon fishing, grouse shooting, stag hunting or tourist season the District officer's Friday arrival for a "surprise check" was not so often a "surprise".

The Commission's work of establishing plantations and carrying out a relatively modest thinning programme in the conifer plantations of the 1920s and 1930s could not provide as much employment as the Government wished in the immediate post-war years of 1946, 1947 and 1948. As the Commissioner's road building programme had lapsed during the war, it was agreed that they should have a crash programme of road building to be undertaken by a new separate Engineer Branch.

Priorities were set by the proximity of forests to areas of unemployment and old fashioned labour intensive methods of road building were given precedence over the now commonly available mechanised methods. This was a thoroughly unsatisfactory, costly way of proceeding and by the end of 1948 the Government agreed to a more sensible approach [**4**].

The Engineer Branch was merged into the general organisation of the Commission and the priority for road building dictated by the needs of the forests for roads for the extraction of thinnings and for access for fencing, draining and planting new areas. Mechanised methods of road building were introduced as rapidly as possible and labour was recruited for its skills, not because it was available and unemployed.

16 Planting Progress to 1949

The 30th CAR [**4**] contains a useful table giving a breakdown of the total land acquired through the Forestry Fund and the Transfer of Woods Act 1923—a total of 1 559 900 acres (631 000 ha)—between 29th November 1919 and 30th September 1949.

"The utilisation or intended utilisation is given in Table 2.6 below, which shows that 949 100 acres are classed as Forest Land, comprising 614 200 acres of plantations and 334 900 acres of land still to be planted. The 334 900 acres still to be planted is well distributed between the three countries: England holds 114 100 acres, Scotland 164 600 acres and Wales 56 200 acres. The distribution is considered reasonably satisfactory having regard to the planting programmes of each country. It should be noted that much of this land is let for grazing as late as possible before it is required for planting".

Table 2.6 Utilisation of land as at 30th September 1949

		GB	England	Scotland	Wales
Total Acquired	ac	1559.9	526.4	820.8	212.7
	ha	631	213	332	86
Forest Land: Total	ac	949.1	394.1	395.4	159.6
	ha	384	159	160	65
Acquired plantations	ac	57.3	41.5	12.1	3.7
	ha	23	17	5	1.5
Planted by FC	ac	556.9	238.5	218.7	99.7
	ha	225	96	88	40
To be planted	ac	334.9	114.1	164.6	56.2
	ha	136	46	67	23
Other land: Total	ac	610.8	132.3	425.4	53.1
	ha	247	53	172	21
Nurseries	ac	2.2	1.0	0.9	0.3
	ha	0.9	0.4	0.4	0.1
Rough grazings and Agricultural land (a)	ac	371.1	62.7	264.8	43.6
	ha	150	25	107	18
Forest Workers Holdings	ac	14.1	6.6	3.3	4.2
	ha	6	2.7	1.3	1.7
Unplantable & Misc.[a]	ac	223.4	62.0	156.4	5.0
	ha	90	25	63	2

All numbers in this table are acres×1000 (ac) or hectares×1000 (ha).
[a]As at 30th September 1947 Agricultural land (62 600 acres) and Unplantable and Miscellaneous (481 800 acres) totalled 544 400 acres. The table above gives as at 30th September 1949, Rough grazings and Agricultural land 371 100 acres and Unplantable and Miscellaneous as 223 400 acres, a total of 594 500 acres.

Figure 2.1. Felling Pine thinnings with bow saw, Thetford 1949. Men working without protective helmets, gloves or other safety measures

References

1. Forestry Commission. *Post-War Forest Policy Report by H.M. Forestry Commissioners.* Cmd 6447 HMSO 1943 Reprinted 1956.
2. Forestry Commission. *Post-War Forest Policy. Private Woodlands Supplementary Report by H.M. Forestry Commissioners.* Cmd 6500. HMSO January 1944, Reprinted 1951
3. 29th CAR
4. 30th CAR
5. 28th CAR
6. Forestry Commission. *Census Report No. 1 Census of Woodlands 1947–1949 Woodlands of Five acres and over.* HMSO 1952
7. Forestry Commission. *Report of the Committee on Marketing of Woodland Produce.* 1956 HMSO
8. Grayson A.J. *Private Forestry Policy in Western Europe*, especially Chapter 18. CAB International, 1993
9. Meiggs, Russell. *Home Timber Production (1939–1945).* Crosby Lockwood, London, 1949
10. Agricultural Sub-Committee of a Cabinet Committee appointed to consider the co-ordination of research work carried out by Government Departments. See 30th CAR p77, 78. Note misprint Forestry Act of 1949 for Forestry Act 1919
11. *The Forestry Commission The First 75 Years* (1994)
12. 37th CAR
13. Dalton Hugh, Memoirs 1945–1960, *High Tide and After.* F. Muller 1962

Chapter 3

The 1950s

1 The 1951 Forestry Act

Under the 1951 Forestry Act [**1**] the Commissioners were charged with "the general duty of promoting the establishment and maintenance in Great Britain of adequate reserves of growing trees." The Commission had become the licensing authority for the felling of timber in January 1950 under a Statutory Instrument made under Defence Regulations. This was now translated into permanent legislation. In carrying out their duties under the Act, the Commissioners were required to consult the newly formed statutory Home Grown Timber Advisory Committee (covering Great Britain) and the Regional Advisory Committees, one in each Commission administrative region, the Conservancy. Committees with similar functions had been in existence prior to 1951 (see below) but it was felt that by making them statutory bodies the Government's intention to consult the private growers and the home timber trade would be more clearly demonstrated.

The Act also required that at least one of the Commissioners should have knowledge and experience of the timber trade.

In order to further strengthen the Commissioners hands in establishing and maintaining the reserves of growing trees, power was given to enable the Commissioners to make the granting of a license to fell trees subject to the land being replanted. Powers were also given—but it was thought that they would seldom be used—to enable the Commissioners to issue directions to owners to fell trees. The idea was that woods that were decrepit and over-mature should not be allowed to stand whilst others better suited as reserves were felled and, perhaps of more practical importance, that there should be no neglect of thinning of woodlands which would produce much needed timber and improve the remaining crop. A felling licence could not be refused for Dedicated woods with an approved Plan of Operations, nor for any woods managed in accordance with a plan of operations formally approved by the Commissioners, nor could replanting conditions be imposed. The reason for these exceptions was of course that in approving management plans submitted by owners, the Commission applied strict criteria in respect of the felling of trees and the restocking of any woods that were approved for felling. Many owners saw the advantages of such forward looking long-term plans, which avoided the risk that they might from time to time be refused a felling license because of some short-term situation. Of more value in the minds of many owners, however, was the fact that Dedicated woodlands were free from possible interference by Local Planning Authorities seeking to impose Tree Preservation Orders under the 1947 Town and Country Planning Act; the dispensation applied to any other woods managed under a plan of operations approved by the Commission e.g. so-called Approved Woodlands (see below).

2 Approved Woodlands

In 1952 the Commission announced the introduction of the Approved Woodlands Scheme (made possible by the 1951 Forestry Act) which embraced estates whose owners decided not to dedicate (some could not for legal reasons) but who undertook to manage them under plans approved by the Commission [**2**].

3 The Home Grown Timber Advisory Committee (HGTAC)

The Home Grown Timber Advisory Committee (HGTAC) is the longest serving and the most influential of all the Forestry Commission's advisory committees and it is perhaps appropriate here to trace its origins and its increasing influence on forestry policy and the implementation of that policy by the Commission.

The HGTAC had its origins in the period leading up to the Second World War because of the problems of meeting the country's needs for timber in the event of war—thought in 1937 to be "virtually inevitable". The Commission was actively engaged with the Board of Trade in planning the harvesting and supply of home timber. The production of sawn timber was to be undertaken by a greatly expanded home sawmilling trade with the Commission acting as a Department of Timber Control to produce mining timber, especially pit props. To get advice from and the co-operation of the trade and private woodland owners (who at that time owned most of the woods with trees big enough to produce pitprops and logs for sawmilling) the Commission set up the HGTAC in 1939 [**3,4**] representing the home timber trade, private woodland owners, the Board of Trade and the Commission itself. One of its first duties was to draw up a schedule of maximum prices for standing timber and round timber in the log, sawn lumber and mining timber and to draft orders for the control of the felling of timber in the event of war.

During the war, the HGTAC was largely superseded by a number of other committees which were set up to assist the Departments responsible for the production of home-grown timber. But whilst these other committees subsequently went out of being with the ending of the emergency arrangements for the production of timber, the HGTAC survived and continued to act in the post war years as a medium for the discussion of felling licensing. In 1950, its chairman was the Timber Controller of the Board of Trade. In January of that year, when the Forestry Commission took over the control of felling from the Board of Trade—still operating under the emergency war-time regulations—the chairmanship of the HGTAC passed to the Forestry Commission. When, as noted above, the Government decided that the control of felling in Britain's much depleted private woodlands must continue indefinitely and the Forestry Act 1951 included a section making the control of the felling of timber in Great Britain a duty of the Forestry Commissioners and made the HGTAC a statutory body, the Commissioners noted in their 32nd Annual Report (1951) that this step had been universally welcomed as proof of the Government's intention to bring the relevant organisations into consultation on all major forestry problems. The Act also made the 11 Regional Advisory Committees statutory bodies.

The 36th CAR (1955) in a six line entry recorded that at the four meetings of the HGTAC "normal business in connection with the Forestry Act 1951" was carried out and that "In addition the meetings provided the welcome opportunity for frank discussion on a number of other matters affecting private forestry and the home-grown timber trade" [**5**].

As is recorded later in this chapter, the Committee on Marketing of Woodland Produce (The Watson Committee) [**4**] were not at all impressed by the "frank discussions" with no formal tendering of advice by the HGTAC to the Commissioners. They felt that the Commissioners should be absolutely open about the advice tendered and the actions taken (if any) by them to implement that advice. In spite of behind the scenes lobbying of the Watson Committee to maintain the status quo, the Committee in its Report, published in 1956, firmly recommended an overhaul of the HGTAC's terms of reference, and was adamant that the role of the HGTAC should be greatly strengthened. Thus its advice would be given formally to the Commissioners; it should have an independent Chairman; its membership should be amended to allow for three independent members with no affiliation to private forestry or the timber trade to be appointed; and for the Utilisation Committee (see below) with its technical secretariat to be strengthened by becoming a sub-committee of the statutory HGTAC rather than a non-statutory Forestry Commission Advisory Committee.

The other recommendations of the Watson Committee are given below under Marketing—The Watson Committee.

4 The Advisory Committee on the Utilisation of Home Grown Timber

On 3rd May 1950, in London, the first meeting took place of a Committee which the Commissioners had appointed in December 1949 "To advise the Commission on measures designed to promote the utilisation and sale of produce from British woodlands" [**6**]. The membership of the Committee comprised the Chairman of the Forestry Commission, the Director General, his deputy and the Director of Research; a representative from the Board of Trade, the Forest Products Research Laboratory, the Rural Industries Bureau, the Timber Development Association and the Home Timber Merchants Association of Scotland; two representatives from the Federated Home Timber Association (which covered England and Wales) and three from the growers organisation, the UK Forestry Committee.

In opening the meeting, the Chairman drew attention to the need to make the best use of the relatively small stocks of mature timber growing in Britain and to the fact that the out-turn of thinnings from the plantations created in the inter-war years was increasing and could in time make a significant contribution to Britain's softwood needs. The Committee could make an important contribution to the two aspects of utilisation. A small staff would be recruited to undertake such work as the Committee and Commission might agree upon.

During subsequent meetings, the Committee identified the main problems as:

1. Finding markets for scrub and small-sized hardwoods including coppice and underwood.
2. Finding markets for small-sized conifer thinnings.
3. The utilisation of sawmill waste.

The utilisation development section undertook market surveys and ad hoc research work. The market research surveys of the next decade included wood flour, charcoal, wood wool, turned wood goods; the use of wood in river and sea defence work; in packaging; in materials handling; and in house construction. The ad hoc research work included sponsored research into the pulping properties of hazel coppice and other hardwoods—which led to the building of a bleached hardwood pulpmill in South Wales. The economics of small-scale wood-processing plants (pulp, fibrewood and particle board) were also studied.

A long-term programme of research into the properties of British timbers at the Forest Products Research Laboratory at Princes Risborough was also launched.

Other important aspects of the Committee's work included taking up cases of unfair discrimination against home grown timber; trying to ensure that codes of practice and official "standards" and specifications (which were mostly based on the assumption that the timber would come from overseas) were not drafted in such a way as to exclude home-grown timber; and the design and erection of prototype houses and buildings in home grown timber by way of demonstration units and long-term service trials.

5 The First Seeds of Change in Forest Working Methods and in Labour Requirements

There was, in the late 1940s, a realisation world-wide that forest work from the tree-planting to the thinning and felling stages could not continue to utilise traditional manual working techniques; that accidents and ill health due to poor training and bad working practices were at unacceptable levels socially and in the long run were costly; and that the growing demand for wood and wood products of all kinds could not be met without improving the productivity of all forest operations—particularly of course the harvesting of trees. These questions were first spelt out in detail by FAO's European Forestry Commission (EFC), the Rome-based regional body that covered most of Europe, including the Scandinavian countries and North America; by the (Geneva-based) Timber Committee of the Economic Commission for Europe (ECE), which included the USSR and the Eastern European countries; and by the long established world-wide International Labour Office (ILO) based in Geneva.

In 1951, the first practical initiative was taken [**7**] at the suggestion of the Timber Committee of ECE, by the European Forestry Commission in collaboration with ILO when it set up an Pilot Committee consisting of eight countries to advise on what steps could be taken to foster international collaboration in the field of forest working techniques and training, health and safety of forest workers. The eight countries were: Austria, Finland, France, Germany (then West Germany), Italy, Sweden, Switzerland and the UK, as represented by the Forestry Commission of Great Britain. Following the recommendations of the Pilot Committee and a joint session of the EFC and TC of ECE, a Joint FAO/ECE Committee (working in collaboration with ILO) was set up in 1954 with a membership covering Europe, North America and the USSR and with wide terms of reference covering all aspects of forest working techniques; productivity; mechanisation; and forest workers health, safety, training and standards of living. Although initially mainly concerned with harvesting of timber (= logging), the Committee subsequently covered all forest operations including the establishment of plantations and became one of the most outstandingly successful of United Nations committees. Its work had a major practical impact on operational efficiency including mechanisation and workers health and safety throughout Europe.

With the benefit of hindsight, it is possible to see that there was at this time—the late 1940s and early 1950s—a certain irony in the fact that whilst forestry's demand for labour and its employment opportunities for ex-servicemen was seen as a practical blessing by Governments, the heads of forest services and the forest industries were planning to improve productivity and so reduce their dependence on man-power. Some individuals and some countries were of the opinion that there would be a shortage of labour as more rewarding less physically demanding alternative employment opportunities to forest work would gradually emerge as Europe's industrial capacity recovered from the war years. So they saw the need to improve the forest workers lot and to mechanise forest operations as a necessity to enable forestry to survive and to meet the rapidly increasing demands for wood in all its many forms for the re-building of war-torn Europe and the USSR and to meet peoples' expectations for an improved life style. It has to be said in this context that the devastation of Britain's forest resources during the two wars had not been mirrored in Continental Europe, including Eastern Europe, where the post-war shortage was identified by the Economic Commission for Europe and the European Forestry Commission not as a shortage of timber growing in the forests but as a shortage of the means to harvest it.

In the event, the drive to improve the productivity of manual work and to mechanise forest operations went ahead in Europe quietly and without publicity in the 1950s. Two things helped in this effort. One was the work done in a few countries during the war (e.g. Sweden, Switzerland) to improve the design of hand tools and to train men to use them just as today's golfers and tennis players and other sportsmen are trained to use and improve their natural abilities and are given clubs and racquets of far superior design to those of the inter-war years. The other was the improvements and new inventions of the war which made major changes possible in the design of peace-time machinery such as wheeled and tracked tractors and power saws (=chain saws). The new range of tractors were able to pull specially designed forestry ploughs for drainage and cultivation of soils previously done laboriously by hand or, in some cases, not done at all because of the impracticability of tackling them by hand. By the 1970s tractors could carry over rough terrain on one machine hydraulically operated attachments for felling, branching, de-barking, measuring and cross-cutting trees; by the 1990s tree harvesters were a major means of felling and processing standing trees into saw logs and other roundwood assortments.

6 The Home Timber Trade

If the Forestry Commission was trying to establish British grown timber as a material in its own right in times of peace, it was also trying to encourage the home timber trade [**4**] to process that timber efficiently and to move from the inevitably lower quality standards of war-time to the much more exacting quality requirements of a peace-time economy. Ironically, the tightening of specifications for timber was taking place at the time of great scarcity of good quality British trees, for war-time fellings had creamed the best stands. Both the English timber trade, which handled mainly hardwoods and the softwood-oriented Scottish trade, complained vigorously that the licensing system, the somewhat ad hoc marketing methods of private estates and the generally low prices for sawn wood made it difficult to stay in business and make reasonable profits.

The Forestry Commission, which had mainly small-sized thinnings to sell, let competitively a few long-term (5 year) contracts for standing thinnings in the hope that merchants would be encouraged to settle workers in an area. In some localities the aim was to try to interest firms long established in the mature hardwood business in the harvesting and processing of small-sized softwoods for the coal mines.

In 1950, following the removal of standing timber maximum prices by the Board of Trade, the Forestry Commission, the growers and the timber trade formed a Committee to try to negotiate a National price agreement for mining timber, by far the largest outlet for small-sized thinnings. During protracted negotiations at the Great Britain level, the Scottish merchants reached agreement with the National Coal Board's Scottish Division.

By the end of 1951, agreement had been reached for most home mining timber supplies to English and Welsh collieries. Annual price negotiations of one sort or another between the National Coal Board and the timber trade and growers—including the Forestry Commission—were to continue for many years and were a major annual event in the forestry calendar.

Details of the prices agreed were published by trade associations. The following extracts from the Home Timber Merchants Association of Scotland (HTMAS) for 1965/1966 illustrate the wide range of sizes and prices for the Scottish Division of the National Coal Board. The English and Welsh Divisions had their own specifications and sizes. Cutting the roundwood available to the merchant to the most profitable range of sizes was based on the skill and experience of the workers in those pre-computer days.

NATIONAL COAL BOARD—
SCOTTISH DIVISION

Contract No. HQ/17/65

Schedule of Homegrown Mining Timber Prices

As arranged with The Home Timber Merchants' Association of Scotland to operate for period 1st October 1965 to 30th September 1966.

Table 1.—Peeled Round Pit Props

Item	Top Dia.	Lengths	Prices per 100 Lineal Feet. Mixed Conifers Peeled
1	2½ ins.	All	18/–
2	3 ins.	All	25/6
3	3½ ins.	All	34/7
4	4 ins.	Up to 7 ft.	46/–
5	4 ins.	8 ft.	51/6
6	4 ins.	9 ft.	53/1
7	4 ins.	10 ft.	54/9
8	4½ ins.	Up to 7 ft.	58/2
9	4½ ins.	8 ft.	63/4
10	4½ ins.	9 ft.	65/2
11	4½ ins.	10 ft.	67/–
12	5 ins.	Up to 8 ft.	72/7
13	5 ins.	9 ft.	78/5
continuing to			
52	8 ins	14 ft	199/6

Table 3.—Sawn Pillarwood

Dimensions of Cross Section	Prices per Cubic Foot Delivered to Colliery Bundled All Descriptions Softwood & Hardwood
1. 2 ins. x 2 ins. 2. 2¼ ins. x 2¼ ins.	5/7½
3. 2½ ins. x 2½ ins. 4. 3 ins. x 3 ins.	5/11½

Table 4.—Chocks, Coverboards, Lids & Wedges

Description	Hardwoods including Oak Prices per Cubic Foot Delivered to Colliery
1. Sawn Chocks—4" x 4"	5/4½
2. do. —other sizes up to and inc. 6" x 6"	5/10
3. Edged Coverboards	5/11
4. Lids, und. 2 ins. bundled	5/4½
5. Lids, 2 ins. and over up to 6 ins. width bundled	5/7
6. Lids other sizes	Prices by negotiation

Prices for Coverboards are for mixed widths 4" to 7" and in lengths with ½" margin over or under. Where all one width is specified or exact lengths, chamfering or tapering prices by negotiation.

Table 5.—Softwood Crowns and Sleepers

Dimensions	All Conifers including Larch — Prices per Cubic Foot Delivered to Colliery
1. 4 ins. x 2 ins. and 5 ins. x 2½ ins. not over 6 ft. long	5/5½
2. 4 ins. x 2 ins. and 5 ins. x 2½ ins. over 6 ft. long	5/6½
3. 6 ins. x 3 ins. up to 6 ft.	5/6½
4. 6 ins. x 3 ins. 7 ft. and 8 ft.	5/9½
5. 6 ins. x 3 ins. 9 ft. and over	6/3½
Splits in sizes 1–3—7½d. per cubic foot less Splits in sizes 4 and 5—8½d. per cubic foot less	
6. 7 ins. x 3½ ins., 8 ins. x 4 ins., 9 ins. x 4½ ins. — Up to 8 ft. 7. 7 ins. x 3½ ins., 8 ins. x 4 ins., 9 ins. x 4½ ins. — 8 ft. to 10 ft. 8. 7 ins. x 3½ ins., 8 ins. x 4 ins., 9 ins. x 4½ ins. — Over 10 ft.	Prices by negotiation
9. Sleepers— 8 ins. x 4 ins. with 6 ins. flat face 8 ins. x 4 ins. with 6½ ins. flat face 8 ins. x 5 ins. with 6 ins. flat face 8 ins. x 5 ins. with 6½ ins. flat face	Prices by negotiation

Figure 3.1. Extracts from "Mining Timber Price Lists 1965–66", published by the Home Timber Merchants' Association of Scotland

7 Forestry Co-operatives

"The Commissioners have always held the view that there was great scope for the development of co-operatives schemes in private forestry and have expressed their willingness to assist any schemes which appeared worthy of support. They have also referred the problem to the Regional Advisory Committees and these have given much time to its discussion."

Thus did the Commissioners Annual Report for 1948 state Commission policy on co-operatives [**8**]. The Report also mentions a guarantee made in respect of the Forestry Co-operative Society in Scotland. By 1954, there were six societies actively working in Wales, the Scottish Society was flourishing and expanding as was an English co-operative, Northern Forest Products covering Northern England.

Financial help was either in the form of an out-right grant to help societies in their early years or as a guarantee against loss up to an agreed annual sum.

In 1955, the Commissioners Annual Report [**9**] said rather plaintively "the policy of the Commissioners towards the foundation of Co-operative Societies and their willingness to assist such Societies is well known, but there has been a lull in the year under review and no new societies have been formed."

Many who joined co-operatives did so only half-heartedly and put in the way of the co-operative only their most difficult sales, themselves engaging directly in the more lucrative or easily effected deals. Some landowners, however, encouraged and helped co-operatives by putting some business their way.

However, estate managers were often lukewarm about co-operatives, particularly if they threatened to take over the marketing of timber and to deprive the manager of all or part of any commission he might earn on sales which he himself arranged direct with the timber merchant or the consumer.

Although inevitably some co-operatives folded up, the foundations had been laid of a co-operative movement, parts of which were to survive and to provide members with a much-valued service. In Scotland, the co-operative was to form the Scottish Woodland Owners Association following on the recommendations of the Watson Committee report of 1956.

8 Education

When the Forestry Commission came into being in 1919, there were few facilities for the training of foresters. An early priority was to establish Forester Training Schools to train local supervisors (foresters). Although aimed primarily at its own needs, some 49 out of the 427 that qualified between 1919 and 1939 found jobs in private forestry or in the Colonial Forestry Service.

The Commission continued its policy of providing residential courses for local supervisors and by 1949 five schools were in operation [**10**]. In 1950 there were

269 men in residence doing the two-year foresters course and in that year 128 men completed their training. A few students were from overseas e.g. two from the Cyprus Forest Service.

9 The Progress of Planting

Private planting and re-stocking in 1954 [**11**] with the aid of grants was some 15 700 acres (6354 ha); additionally, about 3400 acres (1376 ha) were planted without grant aid. In 1955 private planting figures [**5**] were 19 600 acres (7932 ha) and 2500 acres (1011 ha) respectively.

By 1955, the Commission's acquisitions of land and its planting had begun to slow down after the peak of 1954 when some 77 000 acres (31 160 ha) were acquired and some 43 000 acres (17 400 ha) were planted and 27 000 acres (10 927 ha) re-stocked.

Not only had the annual rate slowed down [**12**], but overall Commission planting was well below the target of 900 000 acres (360 000 hectares) of new planting and re-stocking that had been set for the decade 1947–1956; at the end of 1956 only 61% (548 000 acres) of the programme had been achieved. Acquisitions at 27% of the programme of 1.85 million acres (740 000 hectares) were even further behind target. The Commissioners said in their 1956 Annual Report that it was not the total size of the programme (of 5 million acres of fully productive woodland in the state and private ownership) that was wrong, but the rate of achieving this target that was set in 1946. The Commissioners also reiterated their policy of buying land only on the open market or by negotiation where land was offered to them direct by owners; compulsory acquisition they said flatly would not achieve their objectives.

To overcome some of the reasons why owners might be reluctant to sell bare hill land for forestry purposes, the Commissioners said that in appropriate cases, they would continue to maintain hill fences against adjacent farmers grazing stock even when the woods were beyond the stage at which straying stock would do damage to trees. They also said that they would burn off the vegetation round the outside of plantations on hill land so as to allow normal hill burning by farmers or grouse moor owners to be carried out in greater safety.

It was not until 1960 [**13**] that new planting—as distinct from replanting in existing woodlands—by the Commission again topped the 40 000 acre (16 188 ha) figure as it had done in 1954. Planting by the private sector —new planting plus re-stocking—continued to increase steadily and by 1960 the rate was well over 35 000 acres (14 160 ha) per annum.

The Commissioners re-iterated their view that:

From an economic and social standpoint, the original intention to provide from home sources approximately one-third of the country's pre-war consumption of wood in all its forms is a realistic objective.

10 The Zuckerman Report

In their 38th Annual Report for the year ending 30th September 1957, the Commissioners referred to their previous year's report in which they expressed the opinion that:

"There is a general recognition throughout the country and in particular by the owners and occupiers of hill land, that in principle a substantial increase in forestry is desirable with the objective of raising the productivity of much of the hill land which is at present not being put to its best use."

The 38th Annual Report continued thus:

"This view has been strongly reinforced in the recently published Zuckerman Report on Forestry, Agriculture and Marginal Land [**14**]. The Commissioners welcome this report because, for the first time, the whole question of the proper use of marginal land in Great Britain has been thoroughly investigated and by an impartial body. "Marginal Land" means, broadly speaking, the rough grazings and hills which are the areas where afforestation must find its place. It is not appropriate to comment in detail on the Report, but the Commissioners would wish to say that they are in full agreement with it. Three points of special importance for forestry must however, be picked out.

First, the conclusion that forestry and agricultural development should be planned as an integrated whole. The balance between these two forms of land use has been upset and largely destroyed over many hundreds of years by the heedless clearing and neglect of forests. This balance should be restored and the Commissioners' programme of afforestation will do much to bring this about. So that this may be most effectively carried out, the allocation of areas for afforestation should be a joint undertaking with agriculture, due regard being given to the effect of forestry on wild life and on the beauty of the countryside.

Secondly, that end uses for forest products, in addition to the traditional ones, should be found. Uses for wood are continually changing and the tendency is increasingly towards pulping and chipping for the making of paper and many forms of board. One of the most interesting of these recent developments is the use of hardwoods for pulping.

Thirdly, the conclusion that "there is probably a stronger economic reason now than there has been hitherto for investment in planting". In reaching this conclusion, the Zuckerman Report examined the basis for our forest policy. As a result of two world wars the objective was to replace and augment the growing stock of timber as a strategic reserve for use in a future war. The Report sees the justification for our forestry programme more in the social and economic benefits it can bring than in its relation to war-time needs.

As was announced in the House of Commons on 29th October 1957, the Government have decided to carry out a thorough review of the bases and objectives of forestry policy, taking full account of all the economic, social and defence factors involved. Some at least of the views expressed in the preceding paragraph may therefore need reconsideration in the light of this review."

11 The 1958 Statement on Forest Policy

Following on the Government's review of forestry policy, on 24th July 1958 the Minister of Agriculture speaking in the House of Commons said that the Commission's planting programme should be fixed for periods of ten years at a time [**15**]. For the ensuing decade, he gave 535 000 acres (216 000 hectares) as the programme of Commission planting. In the first five years, the programme was to be 300 000 acres (121 000 hectares) representing a slight increase on the then current rate of planting. The target of 235 000 acres (95 000 hectares) in the second quinquennium was of course a substantial reduction over the programme envisaged by the Commissioners.

Behind the announcement that planting programmes would be reviewed regularly at ten yearly intervals, lay a fundamental change in the outlook for forestry. No longer was the target to be 5 million acres of productive managed woodland in Great Britain, part private part State. A strategic reserve of growing timber was not a prime defence consideration. Or as the Commissioners put it—with superb understatement—the strategic value of forestry is no longer so strongly emphasised.

It is, however, noteworthy that in practice the rate of new planting by the Commission turned more upon their difficulties in acquiring sufficient land than upon lack of Government financial help.

It is also noteworthy that the private sector were promised increased grants in 1958 subject to certain conditions, one of which was that they should form representative bodies constituted broadly along lines suggested by a Departmental Committee (The Departmental Committee on the Marketing of Woodland Produce [**4**]). Past achievement and the proposed changes are summarised in Table 3.1 below [**15**]

Table 3.1 Forestry Commission planting and programme

Achievement		
1949–1953	Five years total 284 000 acres (114 933 hectares)	Annual average 56 800 acres (23 000 hectares)
1954–1958	Five years total 310 000 acres (125 450 hectares)	Annual average 62 000 acres (25 000 hectares)
Programme		
1959–1963	Five years total 300 000 acres (121 400 hectares)	Annual average 60 000 acres (24 280 hectares)
1964–1968	Five years total 235 000 acres (95 103 hectares)	Annual average (47 000 acres) (19 000 hectares)

12 The Ending of the Felling Quota. Replanting Conditions

The July 1958 Statement [**15**] announced the ending of the private woodlands felling quota which were fixed at the beginning of the year, the amount of felling that would be licensed in the coming year.

The Statement said:

> "The system of felling licences will be continued. In present circumstances, it does not seem necessary to continue to fix a quota for the total annual felling, but the licensing system will be continued though with some relaxations. A Statutory Instrument will be made so that fellings in dedicated woodlands will no longer require a licence.
>
> In addition, licences will, in general, be freely granted for other fellings subject to the existing arrangements for consultation with planning authorities. In order, however, to provide against the undesirable exploitation of woodland areas, the licenses will normally have a condition attached requiring restocking. And licences for thinning or selective felling will not be granted where they would, in effect, permit exploitation without restocking.
>
> The Government recognise the importance for the forestry industry of an efficient home-grown timber trade. They believe that the measures just announced will be welcomed by the trade as well as by woodland owners, and will help both to plan ahead with confidence."

The Forestry Commission commented:

> "One significant change outlined in the Minister's statement is the abolition of the felling quota. The immediate practical effect of this is that—subject to the power of local planning authorities to make Tree Preservation Orders—a private owner can now manage his woodlands with a view to the most economic management of the estate as a whole."

13 Marketing—The Committee on Marketing of Woodland Produce—The Watson Committee

If Government in 1958 had had second thoughts about the need to declare a long term target of 5 million acres of productive woodland in Britain, private owners had earlier become concerned at the prospects of their being able successfully to market the increasing quantities of silviculture thinnings that would be available from the new plantations. They were also concerned at the prospects of finding outlets for the lower qualities of maturing hardwoods that, they alleged, had been left behind when their woodlands were creamed over by Government during the war (a statement not disputed by the Forestry Commission).

In 1954, the then Secretary of State for Scotland and the Minister of Agriculture and Fisheries, acting jointly and after consultation with the Forestry Commissioners appointed an eight-man independent Committee on Marketing of Woodland Produce [**11**]. Mr. (later Sir) Hugh Watson, Deputy Keep of the Signet, was appointed Chairman.

The terms of reference were:

"With the object of promoting confidence and stability and bearing in mind both the output from Commission woodlands and the need to develop markets, to consider what measures might be taken within the home timber industry to improve the arrangements for marketing produce from privately owned woodlands; and to report" (to the Forestry Commissioners).

This was one of the few occasions when the Secretary of State for Scotland and the Minister of Agriculture and Fisheries, who in their personal capacities were the Forestry Ministers, "acted jointly", in accordance with the requirements of the Forestry Act, to give a direction to the Forestry Commission.

The Committee completed its report just two years later in May 1956.

The Committee's conclusions [**4**] were as follows:

Considerations Basic to the Successful Marketing of Woodland Produce

The essence of the problem of the marketing of home-grown timber is two-fold: to integrate the produce from privately owned woodlands with the rapidly growing output from the State woodlands and to integrate the total output from British woodlands into the total pattern of British timber consumption. We regard the following considerations as basic to the solution of this problem.

1. There must be means for ensuring that the produce comes forward in a planned and regular flow.
2. There must be means for ensuring that a fair share of the market is secured by private woodland owners against the competition of the State forests.
3. There must be means for ensuring that any potential surplus of production can be absorbed, whether by transfer to more distant markets or by the creation of new industries locally.
4. There must be a healthy home timber trade capable of dealing with the increasing output of produce and of adapting itself to deal with the changing nature of that output.
5. There must be continual effort and research by all concerned in the industry to overcome prejudice against home-grown timber, to improve its quality and presentation and to develop new uses.
6. There must be means for ensuring that the returns received by woodland owners for the sale of their produce are not only adequate to support the re-stocking and maintenance of their woodlands, but also in due course to provide a reasonable return on the capital represented by the woodlands.

After full and careful consideration, we have come to the conclusion that to attain these objectives, two things are necessary:

(a) a strong and effective association of private woodland owners;
(b) a central consultative body representative of all the principle interests concerned in the marketing of home-grown timber.

Private growers must be effectively organised in their own interests, to ensure that privately owned forest land is used to its maximum productive capacity and that the marketing of the produce from it will be planned and regulated efficiently. The second proposal is necessary in order to ensure, first that the output of timber from privately owned sources is integrated into the total national output, secondly that the latter is absorbed by existing markets and that when necessary new outlets are found and thirdly, that the profitability of timber production is kept continuously under review.

The Committee felt that the "central consultative body representative of all the principal interests concerned in the marketing of home-grown timber" that it saw as the second essential to the orderly marketing of British grown timber, could be created out of the Home Grown Timber Advisory Committee which the Commissioners maintained under the 1951 Forestry Act. But they felt that to be effective, it would have to be more formal in tendering advice to the Commissioners than it had been; that it should have a non-Commission chairman; and that a proportion of its members should be independent of home timber trade or forestry interests. It was also thought desirable that the Commissioners' Advisory Committee on the Utilisation of Home Grown Timber, should become a Technical sub-committee of the re-constituted HGTAC.

The eventual formation of two associations of private woodlands owners and the rejection by the Watson Committee of the idea of a Producers Marketing Board or of a Commodity Commission in favour of the Central Consultative Body are discussed below.

The Watson Committee ended its Report:

Conclusion

We believe that the existence of machinery on the lines we have indicated above would meet what we regard as the two main needs of the home timber industry today; the need for active supervision and planning in the task of finding markets for the growing output of British woodlands, State and private alike, and the need for a reasonable degree of assurance for the financial health of private forestry. The fulfilment of the ideal of partnership between the State and Private enterprise on which British forestry is founded cannot be left to chance; it must be guaranteed by strong and unified organisation of the private interests and by strong independent machinery to hold the balance between those interests and the State.

14 Private Woodland Owners Associations

In the event, the historical differences in organisation between Scottish owners and those in England and Wales and to a lesser extent the somewhat different sizes and types of forestry estate in Scotland and in England and Wales frustrated the formation of one single strong and effective association covering Great Britain as recommended in 1956 by the Watson Committee [**4**]. Instead, Scotland formed its own Scottish Woodland Owners Association (SWOA) based upon the country-wide co-operative then in existence. The Scottish Landowners Federation, although it still takes a special interest in forestry, gave up its leading role in forestry matters to the new Association.

In England and Wales, the Country Landowners Association had been the leading body representing owners forestry interests and the CLA surrendered its forestry role to a new England and Wales Timber Growers Organisation (TGO).

The two new associations, SWOA and TGO came together in 1959 as the Forestry Committee of Great Britain for the purposes of National negotiations with the Forestry Commission, supplanting the so-called UK Forestry Committee the UKFC—which actually covered Great Britain, not the UK [**16**].

> *Note*: The UKFC formed in 1948 by the Country Landowners Association, the Scottish Landowners Federation, the two Royal Forestry Societies (the Royal Scottish and the Royal Forestry Society of England and Wales) had been accepted by Government as the body representing the collective interests of owners of private woodlands; but the members of the Committee were elected by the constituent bodies and not by owners of woodlands per se (the two forestry societies represented forestry, not forest owners; the CLA and SLF represented all landowners rather than landowners with forestry interests). The Watson Committee [**4**] had felt that central negotiations should be conducted by a body whose members owed their allegiance direct to their fellow woodland owners. SWOA and TGO fulfilled this requirement.

15 Central Consultative Body

Recommendations made to the Watson Committee [**4**] to establish a central consultative body did not go as far as some growers wanted. In fact the Watson Committee had been asked by several powerful forestry interests to recommend the setting up of a Producers Marketing Board or a Commodity Commission to take over the marketing of all British grown timber. The main reasons for the Committee rejecting this strongly advocated course of action are summarised below.

One of the main constitutional difficulties that deterred the Committee from accepting the idea of a Producer's Marketing Board—such as those for milk or eggs or potatoes—was the position of the Forestry Commission. The Forestry Commission as a Crown body would be exempt from the legislation necessary to establish a Marketing Board and without the Commission half the productive forest area would be outwith the scheme; practical difficulties would arise if the Commission were to waive its immunity—it would dominate the Board and render it virtually a State Marketing Authority.

Another practical difficulty that was evident to the Committee was the problem of a central marketing board dealing in produce as varied as timber. Marketing Boards had been considered appropriate to a single fairly easily graded commodity such as milk, eggs or potatoes and for a timber board to take possession of timber, grade it and re-distribute it, would be immensely costly in transport, storage and administrative overheads.

Commodity Commissions have, from time to time, been set up under statute to regulate the marketing of various agricultural products and have taken various forms. Had the Watson Committee found that there was a case to justify price support, by Government or by a levy on imported timber, they would have given consideration to the setting up of a Commodity Commission to administer the scheme. The Committee concluded, however, that they had not been presented with evidence to enable them to say that the financial position of the industry warranted price support.

16 Research on Home Grown Timber

Since the 1920s, the Forest Products Research Laboratory of the Department of Scientific and Industrial Research, located at Princes Risborough, had carried out research into the properties of timber imported into Britain on a commercial scale. It was decided as a matter of Government policy that the Laboratory would also serve the needs of the Forestry Commission by undertaking research on the properties of commercial timbers grown in Britain.

The Laboratory had available to it samples of mature British grown timber from the traditional hardwoods (broadleaves) such as oak, beech, ash, elm and sycamore and conifers such as Scots pine, European larch and Norway spruce. A comprehensive handbook [**17**] on the properties of the traditional home-grown hardwoods (broadleaves) and softwoods (conifers) was published in 1941. It was an invaluable book of reference for consumers who had to switch from imported to home-grown timbers during and immediately after the 1939–1945 war. But there had not been available for testing, representative samples of timber from the large new plantations of exotic species—such as Sitka spruce, Japanese larch and Corsican pine, which featured so much in the plantings of the Forestry Commission and of those private owners expanding their forests on bare land.

To remedy this situation a comprehensive enquiry into the properties of home-grown softwoods with priority for the now extensively planted exotics was launched in 1958 [**16**].

17 Research on Machinery for Forest Use (the Mechanical Development Committee)

The 40th CAR (1959) noted the appointment by the Commission of an Advisory Committee on machinery research [**16**] covering the use of new machinery and the development and modification of existing machines for use in forestry. Members included representatives of timber merchants, landowners and engineering institutions. The Committee became known as the Mechanical Development Committee. Suitable "off the peg" forestry machines were seldom available on the market. Agricultural tractors, though cheap, had a very limited role to play in forest conditions whether for ploughing or for timber extraction. They were often under-powered and too delicate for forest work. By contrast, tractors used for earth moving and other civil engineering work were often too heavy, too elaborate and too expensive for forest use, with the exception of a few types of tracked tractor which pulled forest ploughs used for cultivating and draining peat covered and heathland sites essential to the fulfilment of the post-war planting programmes.

Had British forestry provided a large-scale market, then undoubtedly specialist firms would have moved in with special forest machines, but compared to agriculture and civil engineering the forestry market in Britain was, and still is, microscopic. In many other European countries, the forest machinery markets were much larger, but still small compared to agriculture, so that there was also in Europe a lack of machines built specifically for forestry—most were adaptations of agricultural machines. Even in North America, forestry tractors were often based on earth-moving machines designed for the huge civil engineering market. As already noted, this state of affairs had led to a considerable amount of international collaboration between forest services and research organisations working on forest machinery development in Europe, North America and the USSR. Britain, through the Forestry Commission, played an active part, helped by the fact that though she had little forestry tradition, she was a major producer of farm tractors with a long tradition of mechanical skills and knowledge.

18 Forest of Dean Committee

In 1959 the Forestry Commission published the 1958 Report of the Forest of Dean Committee [**18**]. This report gives an account of the history of the Forest of Dean and makes recommendations on the measures desirable and necessary to secure that the administration of the Forest, more particularly as regards the grazing of animals, which may have to be adjusted to modern requirements. For considerations of space and because it was not concerned with forest policy generally, the report is not discussed here.

19 The First Forty Years—The Commissioners Summary of their Achievements 1919–1959

The Forestry Act 1919 "an Act for establishing a Forestry Commission for the UK and promoting afforestation and the production and supply of timber therein, and for purposes in connection therewith" received Royal Assent on 19th August, 1919 and came into effect on the first day of September of that year. Forty years later, at the end of the forest year (30th September) 1959 the Commission summed up its achievements in its 40th Annual Report to Parliament [**16**].

1. An estate of over 2 400 000 acres (971 000 hectares) of which 1 224 000 acres (495 000 hectares) were planted and 316 000 acres (128 000 hectares) awaited planting, making about 1 540 000 acres of forest land (623 000 hectares).
2. Forests in every county in Great Britain except London and Middlesex, from Borgie on the Pentland Firth to Croft Pascoe on the Lizard.
3. A staff of 15 161 of which 12 512 were forest workers.
4. Some 6000 miles of forest road built (in itself an important contribution to the rehabilitation of much long neglected land), and 4600 houses built or acquired.
5. A steady and rapidly increasing yield of timber from new plantations.
6. The drift from the land halted and in some cases reversed where new forests had been created.
7. Seven National Forest Parks established.
8. Most Commission forests open to the public.
9. About 385 000 acres (156 000 hectares) planted by private owners.
10. More than half all privately owned woods being efficiently managed, mostly under a Plan of Operations approved by the Commission.
11. The home timber trade active and alert to new developments.
12. Commission schools have turned out 1778 trained foresters.

The Commissioners were frank in their admission of early mistakes e.g. in the design of plantations in sensitive landscapes but equally robust in their defence of policies and achievements where they thought that their critics were wrong as the following extracts from their 40th Annual Report (1959) show.

20 Extracts from the 40th CAR

20.1 Yield on taxpayers money

The eventual yield on the taxpayer's money invested in the Commission's estate is expected to be of the order of 3%, compound interest. This is in the Commission's view a satisfactory return, although to many it may seem low compared with yields expected from capital invested in commercial enterprises today. But the Commission can by no means always place strictly commercial considerations first—for example, they can normally invest not in the most productive forestry land but in land which is marginal for agriculture: which means that they cannot always use the most productive species.

20.2 Social aspects of forestry

So much for the tangible assets; there are others less obvious. The Commission was set up primarily as an instrument for securing safety in war, by providing a reserve of timber and partly as an insurance against shortage of world supplies. The strategic value of long-term forestry is no longer so strongly emphasised and there is no immediate prospect of shortage of imported timber though some long-term forecasts at least point to a fairly rapid drying up of the exploitable resources of timber overseas. As these aspects have receded, the social aspects of forestry have assumed greater prominence. The Commission have, directly or indirectly, brought men and their families back to the land consistently and on an increasing scale. The numbers directly employed are impressive enough; to them must be added all those who, in increasing numbers as the woods mature, and as the Commission's plantations increase in area, will be employed by timber merchants and wood-using industries and in the ancillary services such as shops, post offices, schools and libraries, which must be provided or enlarged and can be economically maintained when there is a reasonably sized community to support. For the most part this expansion of employment has been and will continue to be in the remote uplands, where until the Commission's arrival there had been a steady exodus of the young and active to the towns, who left behind them an ageing population, derelict homesteads, churches, chapels and schools, even whole hamlets and land going back to a wilderness.

When the population of an area begins to fall, the social services—schools, roads, shops—inevitably begin to be neglected or abandoned; and this in turn accelerates the drift from the land. The increase of population following large-scale planting creates both a need and a justification for improvements and this tends to reverse the process. The most striking examples occur where the Commission has created a new village. For example, to provide a stable labour force for the forest of Inverliever beside Loch Awe in Argyll, a new forest village, now of 47 houses, has been built at Dalavich. In 1908, when there was no forestry activity there, the resident population of this area was 55, of whom only 11 were children under the age of 16. The population is now 318; there are over 125 school children, for whom the County Council has built a new school in the village. This is only one of the more obvious examples; all communities where forestry is undertaken on a large scale benefit in the same way.

20.3 Forestry an alien way of life

In some quarters forestry has been described as an alien activity, an unnatural way of life, for the people of this country. But in many areas, the traditional way of life had gone before the Commission appeared and cannot be replaced. Even if the criticism is a fair one, it ignores the remarkable resilience of our race and its power to adapt itself to changed circumstances. The tenants and workers on upland farms do not themselves find workers in the forests aliens: the communities to which forestry has been added are becoming remarkably close-knit, a process which is naturally enough all the more pronounced where, for example, it is possible to provide village halls to serve as a centre for social activities for all elements in the community and where the Commission are able to help in other ways, for example by providing labour at harvest time. This is an extension of the integration of land use which was described in the Zuckerman Report and is in fact the Commission's policy; the marriage of forestry and pastoral farming to the benefit of both.

Many Commission workers combine forestry and farming: since 1919 the Commission have established over 1000 Forest Workers' Holdings, each of a few acres, where the tenant agrees to carry out for the Commission, and the Commission on its side guarantees, a minimum of 150 days work in each year. Many of these holdings provide striking examples of the success of private enterprise coupled with regular participation in the State under-taking.

20.4 Conifer plantations a blight on the landscape

Another school of thought holds not so much that forestry is an alien activity, as that it bears an alien face: that the Commission have irretrievably ruined many square miles of unspoilt upland by imposing large blocks of commercially managed conifers on land where the natural cover is heather, bracken, moorgrass or scrub. The Commission would agree that there are some few areas, such as the heart of the Lake District, where large-scale conifer plantations, however well planned, will not be acceptable, at least in this generation: but they submit that, outside those exceptional areas, skilfully planned and intelligently managed conifer plantations, on sites where the soil and climate are best suited to conifers—and this described most sites used by the Commission outside Southern England and many within it—can be a positive enhancement to the scenery as soon as they have passed out of the thicket stage, when no plantation—conifer or broadleaved—is beautiful. In particular, they would not agree with the proposition that conifer plantations, if they are planted at all in National Park areas and areas of outstanding natural beauty, can only be regarded as a necessary evil; after all, tourists go to Sweden, to the Black Forest, to the Jura and to many other places where conifers predominate. The Commission admit that there are examples in their earlier plantations of defects such as straight rides or fire lines which unnecessarily ignored natural features. The Commission can have no complaint when criticism is directed to that kind of thing, though they can complain when their use of conifers is attributed to a wilful insensitiveness to natural beauty. There is no truth in the accusation that the Commission are prejudiced against hardwoods and where hardwoods will make a worthwhile crop the Commission will continue to plant them. But the Commission have to make the best use of land available to them, and if hardwoods will not grow well they believe that it would be as great an aesthetic mistake to plant them as it is to disguise modern houses or offices, built of brick, concrete or steel, as Tudor cottages or Georgian mansions. The Commission claim that, as the forests mature, they will be seen for what they are; at once an economic asset and a source of positive pleasure to the eye and spirit.

20.5 National Forest Parks

The forests offer something more than visual pleasure and an economic return. It was very early recognised that, since many acquisitions, in the upland areas at least, include land which is too high or too exposed for tree planting, an opportunity arose to meet the growing pressure of townspeople for access to open country and especially to the hills. Accordingly, it was decided to declare suitable areas as National Forest Parks, where access could be encouraged in co-operation with such bodies as the Ramblers Association and the Youth Hostels Association and could be aided by a modest provision of footpaths and camping sites. The first Park was opened in 1935 in Argyll.

Altogether the seven Parks now in being extend to over 400 000 acres (162 000 ha) of which 180 000 acres (72 800 ha) is under trees. As funds permit, camping sites and shelters, parking places and footpaths are provided and a Guide to each Park gives visitors an indication of the history, natural history and geology.

20.6 Public access to Commission woods

It is not only in the Forest Parks that visitors are welcome: the Commission encourage the public to visit their forests—and here the word "their" has a double meaning—to the greatest possible extent.

20.7 Schools and Forestry

The interest taken by a number of schools and school teachers in forestry is very encouraging and the Commission do all they can to foster it. Some schools have been allocated woodland on which to establish school forests and so follow the work of Cymmer Afan Secondary Modern School at Argoed in the Afan Valley, South Wales, where forestry operations carried out by the children have been associated with traditional subjects in the school curriculum. Recently a class of the Woodhill Primary School at Woolwich set an example to other schools in towns and cities by under-taking a forestry project involving two and a half hours' work a week for a year, the practical work taking place in a local park. There is thus growing up a generation of young people who have come to accept the Commission's forests as places where, quite simply, they can enjoy themselves and in return learn how to protect the forests and help foresters by an understanding of their work.

20.8 Private Forestry

So far the review has covered only the Commission's direct work in building up a publicly-owned forest estate. Their responsibilities do not end there; they are charged with the duty "of promoting forestry" and not solely state forestry. Progress in private forestry, though less evident to the layman, since it cannot be so spectacular to bring back unproductive or devastated woods into production as it is to create woodlands and forests where none existed, has nevertheless been most impressive. Despite lack of funds, labour and housing difficulties, shortage of plants, high taxation, and uncertainty about the future, many private owners have put their woods in order and have undertaken heavy commitments for the future. To have done so is evidence of courage and of faith in forestry.

Progress between the two wars was modest, but even so a total of over 125 000 acres (50 000 ha) was planted. In the second world war, as in the first, private woodland estates made a vital contribution to victory by providing 40 million tons of timber, thus substantially reducing the burden on shipping and more important, saving the lives of seamen. By the end of the war it was calculated that 484 000 acres (some 196 000 ha) had been exploited, either by clear felling or by the removal of all worthwhile timber.

The Commissioners' Reports on Post War Forest Policy (Cmd 6447, HMSO, 1943) [**19**] in which the Dedication Scheme was outlined, postulated a total planting by private owners, in the first decade of 200 000 acres (80 000 ha) at a rate rising to 25 000 acres (10 000 ha) per annum after the first four years. This programme, like that proposed for the Commission, assumed a more rapid recovery from the war than was in fact realised; but the annual planting programme had in fact been surpassed by 1956 and is now running at over 30 000 acres (12 140 ha) per annum. Altogether since the end of the war, over 260 000 acres (105 000 ha) have been planted by private owners. This represents an outstanding contribution to the country's timber resources. More than half of all privately owned woods are now being, or are on the way to being, skilfully managed and the nation has acquired indirectly a most valuable investment in an essential raw material, at a modest cost to the taxpayer.

20.9 Marketing timber from private woodlands

The Commission, as well as private woodland owners, recognise that the great effort which has gone into the rehabilitation of private woods would be largely wasted if their produce could not find a market at prices which represent a fair return on capital; and the problem of marketing—which is common both to the private owners and to the Commission—has assumed greater importance in recent years. It was very thoroughly examined by the Committee on the Marketing of Woodland Produce (the Watson Committee) [**4**] which reported in 1956; and it was in accordance with a recommendation of that Committee that the associations representing private woodland owners have been recast. One of the main objects of the new associations—the Timber Growers Association in England and Wales and the Scottish Woodland Owners Association, joining together as the Forestry Committee of Great Britain—will be to ensure that produce comes forward in a regular flow and that, if necessary, means are created to ensure that any potential surplus of production can be absorbed. This is a major task and one which will mean a heavy call on the energy and enthusiasm of private woodland owners; but the Commission are confident that the challenge will be met.

20.10 Technical Advances—Research and Education

If there has been a revolution in the place that forestry occupies in the national life, there has also been a revolution in forestry itself and the story of expansion told so far in this review must be completed with an account of how that expansion has been made possible.

In 1919 the old traditions of forestry, brought to their highest perfection in the late 18th and early 19th century, had almost gone. During the second half of the 19th century, mainly as a result of growing industrialisation, the old forestry received blows from which it did not recover; its ruin was finally accomplished in the 20th century with wars and heavy taxation, which brought about the collapse and break-up of many of the large estates on which it had always been based. Thus, although the Commission could build on the work of one or two notable pioneers in modern forestry —for example Sir John Stirling Maxwell's work in turf planting and the use of fertilisers at Corrour in Inverness-shire—they had largely to create for themselves a new technique of large-scale afforestation. Today their achievements are acknowledged by foresters throughout the world and even the oldest-established and the largest forest services are able to learn something from this country.

The main credit for the adaptation of new techniques to the new conditions and for the improvisation of new techniques must go to the general body of Foresters and District Officers in the field and the general directing staff. But the two specialist branches, Research and Education, also made a significant contribution. In 1919 the Research Branch consisted of one man; today it numbers 194 and it is recognised as one of the leading forest research establishments in the world.

Since the main bulk of the Commission's needs for professional forest staff could be met by the university schools of forestry, the primary task of the Education branch has been, throughout the Commissioner's history, to train the large numbers of foresters required. The two Forester Training Schools in existence before 1919 were insufficient and three new schools were opened in 1920. Since then 1778 men have qualified as Foresters at the Commission's schools, now four in number; most of them have taken jobs with the Commission, but a number have gone into private forestry in this country and others to overseas forest services and timber companies. In recent years the Branch has also arranged courses—usually of six weeks' duration—for employees on private estates and it has also provided short refresher courses for land owners and agents, local authority staffs, school teachers and others outside the Commission's services.

It would be impossible, in a short review, to give a full account of the technical advances which have arisen in the course of the Commission's work and have in turn made that work possible; all that can be done is to pick out a few characteristic examples. Nursery practice had a long tradition behind it in this country and the Commission's task was to adapt the systems which had been developed in commercial and estate nurseries to the different conditions required for a large-scale state afforestation programme in which the main species used would be conifers.

From the early days of the Commission, it was recognised that drainage and cultivation of the peat-covered and healthland sites, on which so much of the afforestation work was inevitably carried out, was essential for rapid establishment and satisfactory growth, but the necessity to do this by hand placed a severe restriction on the extent of success that could be achieved. A major break-through came just before and during the 1939–1945 war, with the introduction of tracked tractors and the developments of specialised ploughs which made deep draining and intensive cultivation possible. With the help of these machines it is now possible to deal effectively with land which was formerly difficult to treat at all; good crops of trees can be grown on sites which would otherwise yield only a poor scrubby growth. There is still a problem on the poorer sites of NW Scotland. The difficulties met with in attempts at afforestation in the high country of the southern Pennines, where atmospheric pollution aggravates the difficulties caused by exposure on poor soil, have also still to be overcome.

The principle of using plants grown only from selected seed, which has now been generally accepted, has already shown striking results in young beech plantations where the straightness and vigour of the trees is in marked contrast to those formed early in the Commission's history, when less attention was paid to seed origins. Similar, but perhaps less striking, results will follow from the more careful selection of seed now being made for all species and from the tree breeding programme undertaken by the Research Branch. A change of emphasis in the Commission's main effort—from the formation of plantations to their management—began to appear when the first plantations came to the thinning stage.

In the last few years, problems of harvesting have come to the front. By far the greatest part of the yield from our forests comes from thinnings and it has been necessary to look at the thinning from both the silvicultural and marketing points of view. Foresters now realise that as well as being silviculturalists, they are part of a large commercial undertaking and it was significant that when the Commission set up a Work Study Section, their first concern was with the production (harvesting) side of the work. By no means least important of the latest techniques are the first steps that have been taken in experimenting with the use of fertilisers in checked plantations and pole-stage crops.

In fire protection the introduction of radio equipment was a major step forward and on the statistical side the calculation of the expenditure theoretically justifiable on plantations at risk has helped to control costs.

Until the end of the war there was virtually no building of forest roads. Since then, 4000 miles of relatively cheap low-grade forest roads, using, in the more hilly districts, cheap local material—river gravel, morainic gravel and so on—have been built and the Commission are within sight of catching up the arrears of road-making.

20.11 To Sum Up

British forestry has reached the point at which substantial progress has been made in repairing the damage caused by 19th century neglect and the excessive fellings of two wars. The problems of creating and establishing the country's forest estate have largely been solved; the problems of how best to manage it so as to reap a return on the investment, while at the same time preserving the asset so laboriously built up, are now to be faced.

References

1. 32nd CAR (page 5)
2. 33rd CAR
3. 32nd CAR (page 7)
4. *Forestry Commission Report of the Committee on Marketing of Woodland Produce (The Watson Committee)*[1]. 1956 HMSO
5. 36th CAR (page 12)
6. 30th CAR (page 69) and 31st CAR (page 8)
7. United Nations Geneva 1987
   Timber Bulletin Vol XXXIX, No. 3 Special Issue: *The development of forest working techniques and the role played by the Joint FAO/ECE/ILO Committee on Forest Working Techniques and Training of Forest Workers*
8. 29th CAR
9. 36th CAR (page 11)
10. 30th CAR (page 75)
11. 35th CAR
12. 37th CAR
13. 41st CAR
14. *Forestry, Agriculture and Marginal Land.* A Report by the Natural Resources (Technical) Committee HMSO 1957 and 38th CAR
15. 39th CAR
16. 40th CAR
17. Department of Scientific and Industrial Research. Forest Products Research. A handbook of Home-grown Timbers 1941, HMSO
18. Forestry Commission Report of the Forest of Dean Committee 1958. Cmnd 686. HMSO 1959
19. *Post-war Forest Policy.* Cmd 6447. HMSO 1943, reprinted 1956

[1] *Note*: literature from 1958 sometimes refers to the Committee on Marketing of Woodland Produce as the Watson Committee after the name of its Chairman, Sir Hugh Watson, Deputy Keeper of the Signet.

Chapter 4

The 1960s

1 The Marketing and Processing of Home-Grown Timber

The magnitude of the task of finding new markets for conifer thinnings was assessed when forecasts of future production were made by the Forestry Commission.

These forecasts, published in the Report of the Committee on Marketing of Woodland Produce 1956 (the Watson Committee) [**1**] and the actual removals of the coniferous pulpwood, pitwood and other small-sized material are compared in Table 4.1 below.

Table 4.1 Potential and actual yield of coniferous wood

	Thousands of cubic metres under bark			
	1960	1970	1980	1985
Forecast potential yield from silvicultural thinnings of conifers (all sizes), Great Britain	750	1200	2000	[a]
(Index)	(100)	(160)	(267)	
Actual UK removals of small-sized conifers from thinning and clear felling[b]	550	950	940	1420
(Index)	(100)	(173)	(171)	(258)

[a]Forecast for 1985 not made in 1956
[b]Mainly pitprops and pulpwood and particle board material. Source FAO Statistics for the UK [**2**].

The figures show an increase in total potential conifer thinning production in 1980 of 2.5 times the 1960 figure, with an achievement in small-sized conifer production of 2.5 times the 1960 figure by 1985. War-time fellings had creamed off the best broadleaves (hardwoods) and left owners with the problem of disposing of relatively low quality timber trees and large quantities of scrubby growth and coppice for which traditional uses had almost disappeared.

From the perspective of the 1950s, some of the considerations relevant to expanding the markets for British timber were summarised thus in the Report of the Committee on Marketing of Woodland Produce.

2 Extracts from the Report of the Committee on Marketing of Woodland Produce Chapter III HMSO 1956 (the Watson Report) [1]

2.1 Introductory

Wood is one of the principal raw materials in the world. In a highly industrialised country such as Great Britain, wood in one form or another has innumerable uses, both in its natural state and as a raw material for processing by chemical or other means. In striving to achieve the complete utilisation of forest produce the problem which faces grower and timber merchant alike is one of fitting the various qualities of home-grown timber to the most appropriate uses. Within the limits set by soil and climate the grower can set himself the difficult task of producing timber to an exacting specification in the hope of realising a high price or he can less ambitiously aim at producing to one of the more general specifications commanding a somewhat lower price. Whichever course is adopted, and no matter how intensive the management, only a part of the final crop will achieve the desired standards.

Moreover, in the pursuit of the main objective it will have been necessary to remove in the form of thinnings large numbers of trees, many of which by virtue of their size and quality must find a different end use from the final crop. Again, in the conversion of the tree into manufactured timber, it is often the case that only part of it is suitable for a particular use and it is one of the timber merchant's problems to devise some profitable use or uses for the remainder. The range of markets needed, therefore, if the best use is to be made of all the timber grown in this country, is a wide one; moreover the particular markets needed to absorb particular classes of material must exist in the right places and at the right time.

We cannot hope to conclude in the compass of this Report the exhaustive survey which would be necessary in order to determine in detail how the aim of complete utilisation can best be achieved in this country. All we can do is, first, to consider briefly the main uses of timber to which our attention was specifically drawn and the extent to which they are at present, or are capable of being, met by home-grown supplies, and second to examine the principal problems likely to arise in developing the use of home-grown timber.

2.2 Hardwoods

Our attention has been specifically drawn to six markets for home-grown hardwoods, viz: domestic furniture, railway wagon construction and repair timbers, sawn mining timber, fibre building board and chipboard, pulp and minor produce.

2.3 Domestic Furniture

The greatest single demand for hardwoods is for the manufacture of domestic furniture. In 1948 the domestic furniture industry consumed 10 million cubic feet of sawn timber, or one-sixth of the total sawn hardwood consumption for that year. The Furniture Development Council estimated that by 1954 the consumption of hardwoods by the domestic furniture industry had risen to about 17 million cubic feet of sawn wood; they also estimated that had it not been for the substitution of alternative materials this figure might well have been some 2 million cubic feet higher. The principal species of home-grown timber used in the furniture industry are beech, oak and elm. Exactly what proportion of the furniture industry's requirements is now being met from home sources, and what proportion could be supplied if all suitable British hardwoods were in fact converted into furniture specifications, is not known. The tendency has been to assume that only the highest grades of home-grown hardwoods are suitable for furniture manufacture, and we have had conflicting evidence on the economics of preparing and grading timber of furniture quality from lower but technically suitable grades. We have had no precise data on which to base any opinion as to the relative merits of these views. We would simply, therefore, draw attention to three salient points, namely:

1. the large consumption of hardwoods by the domestic furniture industry,
2. the evidence given to us by the industry that it could and would use more home-grown timber if it were prepared to the desired specification,
3. the strongly held belief of certain witnesses that an increased proportion of the present home-grown hardwood supplies could be profitably converted to furniture specifications.

2.4 Railway Wagon Timbers

Since the beginning of the second world war, railway wagon timbers have formed an important outlet for home-grown hardwoods of a quality somewhat lower than that required for furniture. Before the war, only softwoods were used for wagon flooring and sheeting and the use of hardwoods during the war was necessitated by Government policy and by the shortage of suitable softwoods.

2.5 Sawn Mining Timber

The National Coal Board uses large quantities of timber, both hardwood and softwood, for propping and shoring underground. While the bulk of the Board's purchases consists of softwoods, a considerable quantity of hardwood is also used, virtually all of it in sawn form. The Board relies entirely on home-grown timber for its supplies of hardwood sawn mining timber. The specification is easy to meet and allows of the disposal of much hardwood of a quality for which there are only limited alternative markets.

2.6 Fibre Building Board

The fibre building board industry in this country has in the past been built up mainly on the use of softwoods, but latterly some use has been made of limited quantities of hardwoods in combination with softwoods.

2.7 Hardwood Pulp

There are at present no pulp mills in Britain using hardwoods as a principal source of raw material and certainly there is no regular trade in home-grown hardwoods for the manufacture of pulp comparable to that for the manufacture of fibre building board. Nevertheless, our attention has been drawn to the hardwood pulp mill which is to be set up by private enterprise in Monmouthshire, at which an annual consumption of about 36 000 tons of hardwood pulpwood is envisaged (this is described later in this chapter). We have been given to understand that at least one other hardwood pulping project is under consideration.

2.8 Minor Produce

Our attention has been drawn to the importance of sales of minor produce as a source of revenue on a number of private estates. We confine ourselves here to a consideration of produce obtained from coppice, and in particular sweet chestnut and hazel coppice. Most chestnut coppice is cut for the purpose of making cleft chestnut fencing; hazel is cut primarily for wattle hurdles, wattle garden screens and crate rods. Other species of coppice are cut to meet a wide variety of uses. There is no lack of demand for cleft chestnut fencing and hazel garden screens and although there is competition from other types of fence and screen, it is the shortage of underwood craftsmen which limits production, particularly in the case of hazel garden screens. For other types of minor produce, for example pea sticks, bean rods and thatching material, the demand varies greatly as between one part of the country and another and with some notable exceptions, such as for example pottery crate rods, the price the consumer is willing to pay seldom permits of carriage over long distances, so that markets tend to remain local rather than general. The Rural Industries Bureau has estimated that in England and Wales about 30 000 acres of sweet chestnut coppice, 12 000 acres of hazel coppice and 58 000 acres of other coppice species are worked on a regular rotation and provide regular employment for over 2000 underwood craftsmen, including those employed in rural workshops.

2.9 Softwoods General

The demand for softwoods in the UK is very much greater than the demand for hardwoods. The Board of Trade estimated that in 1949 the consumption of sawn softwoods (including boxboards) was in the following order of magnitude: housing and other building; export and food packing; general industry; transport; miscellaneous home and overseas requirements; shipbuilding and repairing.

2.10 Housing and Building

It has been stated in evidence that the inherent quality of most home-grown softwoods is such that if they are properly prepared, seasoned and graded, a considerable portion should be suitable for housing and other building. The proportion likely to be suitable for high-grade joinery timbers is considerably smaller; joinery is, however, a relatively small element in the consumption of softwoods in housing and building.

2.11 Export and Food Packing

In 1949 about one-quarter of total sawn softwood consumption for that year was used for export and food packing. Home production of sawn softwoods in 1954 was only about one-fifth of this figure. The manufacture of boxes and crates, we were told, can be undertaken from a size and quality of timber not suitable for housing-building. Though not all boxes and crates can be made from low-grade timbers, the specification is in general one which could readily be met from home-grown softwoods.

2.12 Railway Sleepers

The Transport Commission consumes annually about 3 million sleepers and about 2 million cubic feet of crossing timbers. During the year 1954 deliveries from home sources totalled 221 000 sleepers and 200 000 cubic feet of crossing timbers, all from Scottish sources. The Transport Commission stated that their past experience of softwoods grown in England and Wales and hardwoods grown throughout the UK had been unsatisfactory. On the other hand, the Department of Scientific and Industrial Research, to whose evidence we shall refer later, did not make this distinction and specifically referred to sleepers as one of the uses for which home-grown Scots pine (the principal species concerned) was suitable.

2.13 Softwood Mining Timber

The National Coal Board, as has already been noted, uses large quantities of softwoods for the purpose of propping and shoring underground. Softwoods may be used either in the round (pit props) or sawn. Virtually all the round pit prop sizes and a considerable proportion of the softwood sawn mining timber sizes can be met from softwood thinnings. Where softwood sawn mining timber sizes are too large to be met from the general run of thinnings they often provide a useful outlet for mature timber of too low a quality for most other uses.

The overall position, so far as softwood mining timber is concerned, is one of decreasing total demand; but with home sources currently supplying only one-third of that demand it is clear that there will be room for many years to come for an increasing intake of home-grown softwoods, provided that they can compete in price and quality and carry the same service as the imported article and that economic arrangements can be made for transport from distant areas. It is necessary, however, to draw attention to the fact that the market in Scotland is now being almost fully supplied from Scottish woodlands and that it may not be economic to transport to English coalfields.

2.14 Wooden Telegraph and Transmission Poles

Wooden telegraph and transmission poles in Britain are invariably of softwood. The principal users are the Post Office and the electricity supply industry. In the immediate pre-war years the Post Office used more than 200 000 wooden poles a year; during the war this figure dropped to about 120 000 poles per annum, including issues made to defence departments. Post-war consumption has been just over 120 000 poles per annum of which 87% were imported. The total purchases of Electricity Boards amounted to 150 000 poles in 1950; this had risen to nearly 200 000 in 1954 and over 240 000 in 1955. For the past few years practically all the poles purchased by Electricity Boards in England and Wales have been imported, while in Scotland a limited use has been made of home-grown poles (about 8% of total purchases in 1950 and 3% in 1955). For Great Britain as a whole the proportion was stated to be 99% imported and 1% home-grown. The British Transport Commission also uses about 8000 poles annually, all from abroad. There is thus a considerable potential market for home-grown poles.

2.15 Pulp

The growing importance of pulp and pulp products as a use for timber can be illustrated by reference to the figures for trade in these products in Europe as a whole. In 1913 pulp, pulpwood and pulp products together accounted for 20% of the total European import trade in wood and its products; by 1937 the percentage had risen to 33% and by 1950 it was 41%. Similarly exports of pulp, pulpwood and pulp products rose from 30% of total European trade in wood and its products in 1913 to 50% in 1950. These changes in the pattern of consumption reflect, first, the increased demand for such items as printing paper and wrapping paper, secondly, the creation of new demands by the development of new pulp products such as rayon, and thirdly, the increasing tendency to use wood pulp products in certain markets where formerly sawn lumber was the traditional material used; for example corrugated cardboard and fibreboard containers in place of wooden boxes, or fibre boards in place of solid boards. These developments have been important in keeping for wood a part of certain markets which might otherwise have gone to other materials. So far as Great Britain is concerned the demand for home-grown timber for the manufacture of wood pulp has in the past been limited by the inadequacy and uncertainty of supplies. The future development of large-scale processing plants such as pulp and board mills will depend upon ensuring adequate supplies of raw material from within an economic transport radius.

Finally, it is clear that the planning of large-scale development of this kind (board and pulp mills) is an intricate matter and that it can be achieved successfully only through the closest consultation and co-operation between all the interests concerned.

2.16 General Observations on Utilisation

This necessarily brief survey of the utilisation of timber in Great Britain prompts the following general observations:

1. The demand for timber and its products is not static; whereas one or two uses, for example, round mining timber, though still substantial, are on the decline, others such as furniture manufacture are increasing, while comparatively new

uses such as pulp may be regarded as being little beyond their infancy in this country.

2. In spite of the competition of alternative materials, the total demand for timber and its products is far in excess of home production, both present and potential.
3. Although the range of uses open to timber in Great Britain is very considerable, there is one class of market—the market for timber as a raw material for pulp and similar products—in which this country is notably deficient in comparison with other countries; and with the growing importance of small-sized thinnings in the output of British woodlands the time is approaching when this deficiency might well handicap the effective utilisation of the country's timber resources.
4. The problem of securing full utilisation is an intricate one and can be successfully overcome only if there is continuing and well-informed consultation between all the interests concerned in the marketing of home-grown timber.

There is one further matter which is of such importance as to require separate consideration. The ability of home timber to displace foreign timber of comparable intrinsic quality over an increasing area of the total market will depend more than anything else upon its achieving a standard of preparation and service equal to that of the imported article.

Consumers attached particular importance to the need for a proper system of grading for home-grown timber. Certain aspects of the quality of sawn timber can be described by the use of grading rules, and this is customary in the imported timber trade; most of the sawn timber imported is bought on this basis. No general system of grading is applied by the home timber trade in this country; representatives of the trade have said that since the proportion of home-grown timber in the higher grades is so small it is more economic to sell on "the mill run". However, a set of draft grading rules applicable to home-grown softwoods has now been drawn up in consultation with the various interests concerned by the Forest Products Research Laboratory, and it is understood that rules for hardwoods will also be formulated. The evidence we received from consumers left us in no doubt that, whatever might be the difficulties of the timber trade in applying these rules at the present stage, their application would facilitate the use of home timber in markets now served by imported timber.

N.B.: The rest of this chapter deals with the successful development of the markets for "pulp and similar products" (paragraph 2 above) and with the growth and development of the traditional home timber trade—of which the sawmilling sector was the main component—to meet the challenges posed by the Watson Committee.

3 New Wood Processing Industries

In the late 1950s, the question of 'economies of scale' was becoming a key factor in determining the size of financially viable pulp mills that could be established.

What to the grower might be a large quantity of thinnings would be a relatively small amount to the paper pulp or fibre board manufacturer. Transporting wood from a wide radius to a central processing plant was costly compared to many other countries on account of a number of factors, including:

- The high cost of rail freight on a rail network lacking suitable timber loading facilities and specialised timber freight wagons.
- The poor quality of many rural road networks and the absence of a motorway system.

Another factor was that productivity was low because of the lack of a forest work-force trained in the use of and equipped with the best in hand tools and (later) power saws, forest tractors and other harvesting machinery. However, with the establishment of a Forestry Commission work study (= operational efficiency) branch and associated machinery development section, this problem was gradually overcome (long before the rail and road transport problems were solved).

The question of the design of viable 'small-scale' pulpmills was explored under the aegis of the Organisation for European Economic Co-operation (based in Paris) since a number of countries—such as France and Italy—had similar interests to Britain.

Britain funded an extension of the main study to try to find out which of the processing industries was likely to give the grower the best return in the long run under British conditions.

The advice of the consultants (Sandwell of Vancouver, Canada) was that small-scale pulpmills were not likely to be economic unless grafted on to existing paper-mills. New fibre board mills (hardboard or insulating board) must have a fairly high capacity and whenever possible be grafted on to an existing wood using industry which could provide a substantial part of its raw material in the form of wood waste (= wood residues in modern parlance). The consultants ranking of the profitability of the industries under British conditions was (1) paper pulp manufacture using where possible a (mechanical) groundwood process rather than a chemical pulping process; (2) fibre-building board mills; (3) particle board (= wood chipboard) mills. A main reason for ranking fibre-building board mills and particle board mills behind pulp mills was that in Scandinavia and forest-rich countries in Europe, such mills had access to large quantities of cheap raw material in the form of wood residues generated by sawmills and large wood-working plants and were therefore able to manufacture and sell fibre board and particle board on world markets cheaply. British board mills relying on thinnings and able to offer only relatively low prices for their wood supplies would find it hard to compete with other users of small roundwood. In passing it may be noted that, emulating practice in forest rich countries with large forest industries, in 1957 at Queensferry (Flintshire) a sawmilling company also in the pitprops, sawn mining timber and wood wool trades, built an insulation board mill to utilise the wood residues from these activities and that part of the conifer thinnings they purchased standing which were not of suitable specification for use as pitprops, sawn mining timber and wood wool. The capacity of the operation was of necessity very small by international standards. But, it had the advantage of being close to many of its customers with whom it could deal direct and to whom it could offer special services.

When it had reached the end of its planned economic life—and international competition had increased—the company did not re-invest in a new plant.

> In their 1959 Annual Report [**3**], the Commission published Sandwell's findings [**4, 5**] and significantly, indicated their willingness to guarantee minimum supplies for new industries at a fair price where that was necessary to enable a start to be made.

It is hardly surprising that over the 40 years that have elapsed since this advice was given, there have been many technological changes in manufacturing methods and major changes in end-product specifications. These have given all wood-processing industries the ability to use less exacting specifications of roundwood and a wider range of wood residues. This had benefitted the grower, the forest products industries and the consumer of the end products, but has still left the paper pulp industry generally able to pay more for its wood than the board industries.

Saw logs have always commanded a higher price than the other forest products industries, with the exception of the veneer and plywood industry, but with higher prices come higher specifications in terms of size and quality, so the grower has to give more in return.

Within these broad generalisations, there are variations as market prices for sawn lumber, paper and board products fluctuate on world or regional scales. Thus, when paper pulp mills are experiencing high demands they can afford to outbid sawmills for the smaller sizes of log that are normally sold to sawmills.

4 Broadleaves (Hardwoods) For Pulp Manufacture

Broadleaved thinnings were forecast to remain static over the period 1960 to 1980 at around 500 000 m^3 per annum in woods of 2 hectares and over. In addition, there were large quantities of small-sized poor quality broadleaves in woods and copses under 2 hectares and in woods that had been creamed and not re-stocked during the war and in neglected coppices. About half the total volume of medium to poorer quality mature broadleaves is in the crown. Traditionally, on felling the main stem, the crown wood (branchwood) was cut into short 4 feet lengths and piled as cordwood in 8 feet long stacks 4 feet wide by 4 feet high. Its main use was for firewood, sometimes for charcoal and, in post-war years, for chocks and cover-boards for use underground in coal mining. Supplies far exceeded demand. Following trials in the Forest of Dean, it was found that worthwhile quantities of pulpwood quality material could be graded out by the wood-cutters and an additional potential use created for this by-product of the felling of large mature broadleaves. So, in addition to finding markets for conifer thinnings, the newly formed Utilisation Development Section of the Forestry Commission was charged with finding uses for all this hotch potch of low-grade broadleaved material which occurred mainly across southern England.

At this time, the manufacture of paper pulp from broadleaves was in its infancy, relative to the use of conifers which formed the major raw material of the industry. However, during the war, British paper-makers, who used almost exclusively imported pulp, had had to use a proportion of broadleaved pulp and had learnt that in limited quantity it could give some papers advantages over those made from 100% conifer pulp.

The Forestry Commission in co-operation with the British Paper and Board Manufacturers Association took the initiative to confirm the technical feasibility of making paper pulp from the broadleaves commonly found in Britain, including hazel [**6**]. Later, Wiggins Teape clinched the matter by sending sample consignments of hardwood pulpwood from the Forest of Dean to Madison, Wisconsin, USA—then the world's leading Forest Products Research Laboratory—for pulp-making trials.

It was necessary to assess the quantities of these unutilised hardwoods growing in different regions in Southern England and estimates were made by the Research Division at Alice Holt Lodge of the quantities of these broadleaves within various radii of a number of centres in Southern England, where there were believed to be adequate water supplies and effluent disposal facilities (the pulpmills of those times used vast quantities of clean water and produced equally large volumes of effluent that had to be discharged directly into the sea or into fast flowing estuaries where it was rapidly diluted). This limited possible sites to places close to the coast, with the disadvantage that their areas of supply of pulpwood were half circles. The only inland exception was around Gloucester, where the Severn estuary gave the necessary fast tidal discharge possibilities. When it was discovered that the Severn railway tunnel had to be kept dry by the pumping night and day of huge quantities of water into the Severn a preferred location was identified at Sudbrook. The pulpmill came into operation in 1958 [**7**].

Not only was Sudbrook a small-scale mill by international standards, but it used a wider range of broadleaved species than any other mill then in operation. When the UK reported on what had been achieved to the European Forestry Commission of FAO, and the Timber Committee of the Economic Commission for Europe, large numbers of visitors came from all over Europe to see what lessons they might learn from this pioneering British achievement.

It was the first new wood-processing industry in Britain to have part of its wood requirements guaranteed by the Forestry Commission under a long term contract. The greater part of its supplies were bought on the open market. A large number of small operators set up to buy standing timber from farmers, private woodland estates and the Forestry Commission and supply Sudbrook with pulpwood.

At that time, many owners were converting their broadleaved scrub woodlands and unworked coppices to conifers. To have a market for the hitherto valueless small-sized broadleaves that had to be cleared prior to replanting with conifers was a godsend. Some owners found that it was financially rewarding to grow a second rotation of coppice for pulpwood for which very little capital expenditure was required, rather than replant with conifers, which meant substantial investment until the new conifer crop could be established.

Wiggins Teape introduced a Hardwood Pulpwood Registration Scheme. Owners who set out to grow a hardwood pulpwood (e.g. old broadleaved scrub or coppice sites) and registered their schemes with the company, were given a guarantee that the pulp mill would, at an agreed future date, buy the pulpwood at the top of the then ruling market price. This novel scheme was more important for its psychological effect on owners than for its influence on the actual future supplies of wood to the mill.

There was a downside to this innovative and important pulpmill. Because the end product was a bleached pulp that had to be free of all blemishes when turned into paper, the bark on the pulpwood billets had to be completely removed before they were chipped and fed into the digesters. From the forester's point of view, this meant that it was not always easy to meet the strict straightness criteria required by the de-barkers. From the company's point of view, it meant careful scrutiny of every load of pulpwood entering the wood-yard.

In time, the pulpmill was modified to produce an unbleached pulp that did not require the pulpwood to be de-barked and which was used to make corrugated paper for packaging. For this product, broadleaves had an advantage over conifers in that they gave the corrugated paper more strength than conifers in certain key aspects. And the fact that the pulpwood did not have to be de-barked in machinery which was demanding as to the straightness, shape and diameter of the pulpwood billets meant that a much higher proportion of the tree could meet the new relaxed specification.

5 The Development of the Traditional Home-Grown Timber Trade

Whilst the need for new wood-processing industries was recognised, the Commissioners were well aware of the key importance of the traditional home timber trade to the prosperity of British forestry. Put simply, the highest prices obtained for timber—hardwoods and softwoods—were when they were converted by sawmills into sawn lumber or plywood; as against this logs could only be sawn profitably if they were above a certain diameter and met stringent straightness and other criteria.

The tops of trees from which sawlogs had been cut and thinnings below sawlog size, producing in aggregate a great deal of wood, had to find other outlets e.g. as pitwood and pulpwood. And critical to the profitability of softwood sawnmills was the ability to sell the "waste" = residues arising during the squaring of the logs and of the deals, battens and boards.

The war years 1939–1945 had left a sawmilling industry geared to the needs of war. To encourage new investment the Commission set out to establish a level playing field to give each and every merchant an equal opportunity to modernise and expand. One key element of that marketing strategy was to develop standards to govern Commission marketing practices in all its many offices scattered throughout the country.

The need for these codes of practice were not confined to the marketing of Commission timber. In the 1950s and 60s, the different branches of the Forestry Commission had developed and published for use by staff throughout the Commission, various Codes of Practice which brought together in a structured way the guidance and the rules of conduct which had been developed over the years, since the Commission's inception in 1919.

The Accounting Officer responsible to the Government for the proper use of the Forestry Funds allocated to the Commissioners by the Treasury under the Forestry Acts was the Director-General. To advise him, he had a Controller of Finance—a professional accountant with a small specialist team under him. Together they laid down the ground rules on which the different branches based their codes of practice.

It would be tedious and serve no useful purpose to go over all the Codes of Practice (or their equivalents) in use in the Commission. Some points from that which governed the Sales of Produce may serve to illustrate the principles which the Code sought to translate into day-to-day guidance for the staff and scrutiny by the internal and external auditors who from time to time carried out very detailed examinations of the accounts at the different levels of administration. Over time, the management structure of the Commission changed—generally to reduce costs by reducing the number of levels of management. But in principle the Chief Marketing Officer at Headquarters, in co-operation with the Controller of Finance, was responsible for preparing and issuing the Code to the next level—the Conservancy of which there were 11 covering the country in the 1960s. The Conservancies, acting through their District Managers gave guidance and instruction to the foresters in charge of the Forests, the basic unit of administration.

An essential function of the Produce Code was staff training, so that the Code covered, for example, the philosophy underlying the Commission's marketing strategy, the broad methods to be adopted in achieving the main objectives of marketing strategy and the nuts and bolts of implementation. It is with the first two aspects that are of our concern here.

The theme running through the Code was that in all its dealings, the Commission must be fair and be seen to be fair to all those purchasing or seeking to purchase

Commission standing timber or felled produce. At the same time, the Commission must seek to obtain the best price possible taking the long term view. It must establish confidence within the timber trade and so attract the large investment needed by it.

The home timber trade were mainly sawmillers, but included round timber merchants who bought and felled standing timber—especially broadleaves—and sorted it into species and sizes of logs for sale to individual customers in the species and qualities they preferred to buy.

The Forestry Commission itself acted as a round timber merchant handling mostly conifer thinnings.

Many sawmills were in need of modernisation requiring substantial capital investment. Whilst the hardwood trade were short of good quality sawlog material—due to the war time plundering of broadleaved woodlands—the softwood trade were faced with increasing volumes of small-sized thinnings requiring a different type of sawmill from the traditional log sizes. Material too small to saw ("small roundwood") found its main outlet as the mining timbers used by the National Coal Board. The new wood-processing industries in the pulp, fibre and particle board fields were coming on stream; depending more and more on both the Commission and timber merchants for supplies; they provided a welcome additional market for small roundwood and sawmill residues.

The Code examined the options for the sale of standing timber—one of the main methods of sale by the Commission. Sale by negotiation was not thought to obtain the best price—and in time this view was confirmed from an analysis of the Commissions records of hundreds of sales over the years—comparing like with like so far as was practicable. Sale by sealed competitive tender brought higher prices, but tended to cause uncertainty in the market—a merchant long established in an area might find himself with no timber to work for the sake of a small amount of money bid perhaps by a rival from another region—who might (and sometimes did) find he had paid over the odds. Sale by auction had the principle advantage of openness. A merchant with good reasons to remain in an area could judge what he had to pay to do so and decide if the cake was worth the candle. Prices paid at auction were published in full. Merchants could therefore build up their own data bank of prices ruling for different species and sizes of standing timber throughout the country and on this base their own company strategies.

The Commission, with the agreement of the home timber trade, developed a method of sale of standing timber that offered continuity to an individual timber merchant for part of his requirements of round timber, leaving him to compete on the open market for the rest of (and often greater part of) his requirements. Thus parcels of standing timber (mainly conifer thinnings) were sold by competitive tender or at auction with a clear understanding that the successful purchaser could buy by negotiation in the same forest a stated quantity in later years. Later a more sophisticated variation was based on a yearly price index calculated and published by the Commission which showed how prices were moving and were used to set the level of increases or decreases on the next year's purchase. The price index calculated and published by the Commission was based on sales by tender and by auction. By agreement with the timber merchants associations (one for Scotland, one for England and Wales) the calculations were checked by their Secretaries who were pledged not to reveal the details of the individual sales by tender.

Timber measurement, particularly of broadleaves, is as much an art as a science as the records and anecdotal evidence from the ancient royal forests of the Dean and the New testify. The published work of the Forestry Commission's Mensuration Branch—especially that based on its innovative work on measuring standing trees—laid the foundation for a more objective approach and came to be accepted by buyer and seller.

Standing trees measured for sale continue to grow until they are felled and rules were laid down for calculating how much they had grown during each season between measuring and felling and therefore what price increase was justified. Mostly felling takes place so soon after measurement that the original volume is used as a basis for payment. But to cater for delays of a year or more (which includes one or more growing seasons) tables were produced giving the percentage increase in volume to be applied in amending the original price upwards.

Much of the rest of the code dealt with matters common to any business such as how to treat purchasers who failed to pay on time, failure by the seller to perform his side of the contract and disputes centred round these two possible events.

As already mentioned above, the main object of the code was to provide a level playing field for all so that the Commission's customers could expect to be treated in exactly the same way from Caithness to Cornwall and would be encouraged to invest in new equipment and to expand their business in line with the increase in availability of timber from the State and private plantations established under the positive Government forest policies.

As will be seen the policy of offering some measure of continuity of supplies to the home timber trade paid off in the long run. As supplies of coniferous sawlogs increased, the more progressive merchants invested large sums of money in new sawmills able to compete with the most up to date mills overseas and to offer customers accurately sawn, graded and seasoned (dried) timber.

6 The 1963 Policy Statement

In 1963, the planting programme was reviewed as had been promised in 1958. On 24th July in both the House of Commons and the House of Lords, there was a Ministerial Statement on forestry policy [**8**].

The emphasis on new Commission planting was to be in Scotland and Wales because there the expansion of forestry was reckoned to bring considerable social and employment benefits.

The Commission was to bear in mind in future programmes the need to provide public access and recreation and to devote more attention to increasing the beauty of the landscape. The role of private planting was recognised and grants were to be continued confirming "the Government's confidence in private forestry" and providing support for "its continued place in our national forest policy." The statement also confirmed the importance of finding adequate markets for home produced timber and the benefit to the balance of payments from the increasing production of wood at home.

The actual planting target for the decade 1964 to 1973 was set by Ministers at 450 000 acres (180 000 hectares) for the Forestry Commission, compared to 609 000 acres (246 000 hectares) actually planted in the decade 1954 to 1963 (private owners had planted 294 060 acres (119 000 hectares), an average of over 29 000 acres (12 000 hectares) per annum in the same period, most of it with grant aid).

The Commissioners response to Government was that large tracts of lands in the uplands where populations were fast declining were indeed very suitable for afforestation and that they would there, in co-operation with the Agricultural Departments, concentrate most of their efforts to acquire plantable land.

It was also in 1963 that an Act [**9**] was passed to empower the Government by way of grants and loans to aid in the establishment of a pulp and paper mill at Fort William in the Western Highlands of Scotland. The establishment of the mill, with all its implications for stimulating employment and the rehabilitation of public services (e.g. postal, schools, housing, roads) was possible only because the Commission was able to guarantee a proportion of the wood supplies for the pulp mill.

It was indeed a bold and imaginative step to set up such an enterprise in such a remote region where hitherto—apart from the aluminium smelter with its own abundant hydro-electric power supply—attempts to encourage new industry of any kind had failed totally.

The Commissioners summed up their policy in their 44th Annual Report (year ending September 1963) to Parliament as follows:

1. To increase the production of wood as a raw material for industry by extending the area of their forests at a steady rate, in accordance with sound land use and by making each forest as productive as possible.
2. Within the limits set by their other objectives to manage the forest estate as profitably as possible.
3. To provide employment in rural areas, especially those most affected by depopulation and in so doing to maintain a skilled labour force.
4. To help in maintaining an efficient home timber trade.
5. To give due attention to the aesthetic and protective roles of the forest and to encourage open-air recreation.
6. To foster industrial and social development ancillary to forestry.

7. To encourage the orderly development of private forestry and specifically to assist in creating conditions in which produce from private as well as Commission forests can be marketed to best advantage.

To give point to their undertakings under paragraph 5, the Commission announced the appointment of a Wild Life Officer, who was to try to improve wildlife management and forest protection throughout the Commission in liaison with the Nature Conservancy Council and other bodies.

The Commission also drew attention to the fact that it was taking advice from a leading landscape architect on the layout of plantations and the conduct of forest operations with an impact on the landscape.

7 Changes in Commission Management Structure

In 1965, following recommendations made by the Estimates Committee of the House of Commons in their Seventh Report (Session 1963–1964, The Forestry Commission, HMSO) [**10**] the Directorate offices in England, Scotland and Wales were abolished and the regional offices (i.e. Conservancy offices) came directly under Commission Headquarters [**11**]. The Director General was appointed a Commissioner and three full-time Commissioners were also appointed covering Administration and Finance; Forest and Estate Management; and Harvesting and Marketing respectively. These changes were designed partially to bring the Commission into line with current practice for bodies with similar commercial functions by cutting out one level of management and giving the chief executive (the Director General) the services of three functional full-time specialist Commissioners, who in turn were supported by specialist staff.

The eleven regional (Conservancy) offices in Britain were also re-organised on functional lines. As a consequence of the abolition of the three national Directorates, the three National Committees which had exercised a measure of executive authority were given new terms of reference. From now on, the Committees were to have a purely advisory role, maintaining liaison between the Commissioners and the specific interests of the countries they represented.

8 Education and Training—The Forester Training Schools Closed

Since 1919, the Forestry Commission had trained in its own schools young men wishing to qualify as Foresters, for employment in both the Commission and private forestry. In 1969, training was taking place at three Forester Training Schools; at Parkend in the Forest of Dean, Faskally near Pitlochry in Scotland and Gwydyr in North Wales. In that year, the Commission agreed to begin to hand over forester training to the public education system, so that forestry training would be on the same basis as, for example, agriculture and would comprise a three-year course, including a sandwich year in a Forest leading to an examination for an Ordinary National Diploma in Forestry.

9 Training of Supervisors and Forest Workers

Training schemes for supervisors and for workers were by 1969 a regular feature of the Commission's training. Some 25 courses were attended by 200 Foresters and 2000 Commission workers attended 500 training courses of about one week's duration given by the Commission cadre of training instructors throughout the country [**12**].

10 The Countryside Acts

The Countryside Acts, Scotland (1967) and England and Wales (1968), gave the Commission specific powers to provide facilities for recreation. The 1968 Act also gave the Commission specific powers to plant and to manage trees in Great Britain in the interests of amenity [**13**].

11 The Forestry Act 1967

This Act consolidated the Forestry Acts from 1919 to 1963 [**14**].

12 Marketing

In 1963 the Commissioners signalled an important change in emphasis from the planting programme which had received most attention to an equally pressing objective—to market efficiently the growing volume of produce. At the risk of duplicating some of the information given elsewhere, it may be worth referring here to some of the main points made by the Commissioners in their 44th Annual Report [**8**].

The increase in production of home-grown softwoods which will rise even more steeply from about 1967 has coincided with the fall in the use of timber in the mines (mines in this context means coal mines). The importance of the mining timber market can be gauged from the fact that in recent years about one-third of Great Britain's total forest production has gone to the National Coal Board. The consumption of home timber by the mines is tending to rise slightly but in the long term it cannot be expected that the mines will absorb a large proportion of the increasing production.

In this situation, it is essential to encourage the planned development of a diversity of wood-using industries based on home timber to provide better and expanding markets for the whole of the country's potential production. The problem has been and will continue to be tackled in a number of ways:

- By removing the prejudices against home-grown timber which is largely a legacy of the (1939–1945) war and the early post-war years when inevitably much indifferent material had to be used.
- By a constant effort by timber merchants and growers to improve standards of preparation and continuity of supply.
- By exploring, with the aid of the Forests Products Research Laboratory of the Department of Scientific and Industrial Research and the Timber Research and Development Association, new ways of preparing and using home-grown timber.
- By improved methods of marketing.
- And, perhaps most important of all, by making it possible for industrial interests to invest in new wood-using industries or the expansion of existing ones. Such investment involves large expenditure and cannot usually be undertaken without assurances of continued supplies of raw material. At present, the Commission alone are able to give such assurances, and they will continue to give guarantees of supplies for new projects requiring substantial capital investment where, in their view, it is in the interests of forestry to do so. In doing this the Commission will continue to take fully into account the interests of private growers and of the trade (home-grown timber merchants) and will not seek to guarantee supplies either in greater quantity or for a longer period than is necessary for the establishment of the new project.

13 Ministerial Responsibility for Forestry

Ministerial responsibility for forestry in England and Wales was subject to several changes between October 1964 and February 1967, when, finally, the Government decided on the Secretary of State for Wales and the Minister of Agriculture for England. Throughout the Secretary of State for Scotland remained its forestry minister [**14, 15**].

14 The Successful Sixties—A Summary

As noted above, Marketing of British timber became in the 1960s an "equally pressing objective" [**8**] to the planting programme. Nevertheless, Government policies promoting the extension of the forest area of Great Britain were very successful in the 1960s. New planting rose from a total of over 78 000 acres (31 500 hectares) in 1960 to over 93 000 (38 000 hectares) in 1970, almost equally divided between the private sector and the Commission. Before the Commissioners acquired land for planting, they sought the agreement of the appropriate Agriculture Department to the proposed change of land use from agriculture to forestry. This was in line with the Commissioners practice and policy first adopted in the mid-1940s.

Private planting on agricultural land required no such clearance and the Forestry Commission had no remit to refuse grants in cases where they or the Agricultural Departments felt that the land should not be planted. Since the Government had given the Commission a target planting programme, the Agricultural Department in Scotland, where almost all Commission acquisitions were taking place, gave their blessing to the Commission acquiring some land that in the normal course of events they would have preferred to remain in agriculture. The Commission and private planters were of course bidding on the open market and it could be said that the failure of sheep farmers to be able to compete for some 80 000 to 90 000 acres (33 000–36 000 hectares) of land a year was no more than an indication of its low value for sheep.

Sheep farmers were divided in their attitude to afforestation on this scale. Of some it was said that they had felt able to continue in sheep farming only because they knew that in the event of financial failure they could sell their land for forestry and make enough money to clear their debts and have some capital for retirement or a new business venture. Others felt strongly that the "subsidy" element to private forestry through taxation and grants and the Commissioner's access to large amounts of public money put them in a position to pay more for land than its "intrinsic worth".

Others argued that when land is bought and sold competitively on the open market, the price obtained does in fact establish its "intrinsic worth". Whatever the merits of the various arguments, the fact remains that the officers of the agricultural departments (especially in Scotland and Wales) often found themselves in an impossible position, knowing that if they asked the Commissioners not to acquire a sheep farm for planting, it might very well go to a forestry investment company, yet finding it difficult to defend on land use grounds their agreement to its acquisition by the Forestry Commission.

It was to be some years before grant aid for planting by private owners was to be subjected to similar clearance procedures by the Agricultural Departments as was Forestry Commission planting.

References

1. Forestry Commission. *Report of the Committee on Marketing of Woodland Produce* (*The Watson Committee*) 1956. HMSO
2. *Timber DATABASE.* 1964–1991; Food and Agriculture Organisation; FAO, Rome
3. 40th CAR
4. Forestry Commission. *Small Pulp Mill Survey 1959.* HMSO
5. Forestry Commission. *Board Mill Survey 1959.* HMSO
6. Forestry Commission Bulletin No. 27. *Utilisation of Hazel Coppice 1956.* HMSO

7. 39th CAR
8. 44th CAR
9. The Fort William Pulp and Paper Mills Act. May 1963
10. Seventh Report from the Estimates Committee (of the House of Commons). Session 1963–1964. The Forestry Commission. HMSO
11. 46th CAR
12. 50th CAR
13. 49th CAR
14. 48th CAR
15. 45th CAR.

Chapter 5

The 1970s

1 Review of Forestry Policy

In 1970 the Government decided to review their forestry policy; in the 51st CAR for 1970–1971 [**1**] the Commissioners commented:

> On 8th December, 1970, the Parliamentary Secretary, Ministry of Agriculture, Fisheries and Food, announced that the Forestry Ministers were to review various aspects of forestry policy, including the return on the public money invested in forestry. It was subsequently announced that the review, which would include Cost/Benefit Studies, would relate to both the public

and private sectors.

The Commissioners welcome this review which should result in clarification of their own terms of reference and the role of private forestry.

It is not always recognised by the public that the Commission is not an independent body. It is no more and no less than an instrument of Government policy, like any other Department of the Crown, and as such it looks to its three Ministers for its instructions on its major objectives and the size of its programme. The policy review is therefore of paramount importance to the future of the Commission.

2 The New Forest

In 1970 the Commissioners set up a New Forest Consultative Panel [**1**] to provide for regular consultation between local opinion and the Commission on "the factors which need to be taken into account in the Management of the Forest".

In May 1971, the Minister for Agriculture in consultation with the Forestry Commissioners issued a mandate for the management of the New Forest, which gave greater priority to the conservation of the traditional character of the Forest as a National heritage and correspondingly less emphasis on timber production, especially in the Ancient and Ornamental Woodlands.

3 The Forest Policy Statement of June 1972

In June 1972 [**2**] the three Forestry Ministers referred to their discussion paper on Forest Policy [**3**]. The paper noted the changes in the Commission's financial objectives, gave a brief summary of a Cost/Benefit Study [**4**] carried out by a group of Government economists and, pending any final decisions that might be reached, put an end to all grant-aid schemes for private woodlands except the legally binding Dedication Schemes already in existence. The Dedication Schemes were however closed to new entrants. As Table 5.1 shows, new planting in private woodlands fell rapidly and by 1976–1981 was less than half the 1971–1975 figures.

The findings of the Cost/Benefit Study, which was the work mainly of Treasury and Forestry Commission officials, was that new planting could not be justified by direct and quantifiable benefits alone. The paper concluded that the only grounds on which further new planting could be justified was the creation or maintenance of employment in rural areas, which would otherwise be substantially depopulated. Even here, the paper suggested, the creation of new woodlands was an expensive way of creating jobs. However, the paper admitted that no other practical way of maintaining rural employment in remote areas had been suggested.

The paper proposed that future new planting by the Commission should be below the target of 60 000 acres (24 000 hectares) per annum put forward in 1967

Table 5.1 New planting (afforestation) by the Forestry Commission and by private woodland owners in ha×1000 in years ending 31st March (60th CAR)

Year	FC	PW
1971	23.0	18.9
1972	21.7	19.7
1973	19.4	20.0
1974	18.4	18.9
1975	19.7	18.3
1976	17.2	9.6
1977	15.5	7.2
1978	14.1	6.3
1979	11.8	7.9
1980	15.8	8.3

N.B.: PW figures exclude new planting without grant aid, averaging about 500 ha per annum in the 10-year period (for which the official estimates do not distinguish between new planting and restocking)

for achievement by 1976. The paper pointed out that in any case the difficulties of acquiring land was making the 60 000 acre target quite unrealistic.

The figure proposed in the policy paper was 55 000 acres (22 000 hectares) per annum of new planting and re-stocking combined; with re-stocking then running at around 10 000 acres (4000 hectares) per annum, this implied a new planting programme of 45 000 acres (18 000 hectares) falling away as felling and therefore re-stocking of older Commission woodlands increased. Thus by 1990, new planting was envisaged to be only 35 000 acres (14 000 hectares) per annum with a further drop to 25 000 acres (10 000 hectares) per annum by year 2000. The programme was to be revised every three years when the relative advantages of forestry were to be considered along-side alternative investment possibilities such as new small industries. The Commission was also asked to pay more attention to enhancing the appearance and attraction of its woodlands "despite some reduction in profitability"; more emphasis should be given to realising the recreational potential of State Forests.

The re-introduction of grants for private planting was to be conditional on the establishment of the woodlands creating employment in areas of rural depopulation or creating or preserving amenity. The grants were to be on a once and for all basis to simplify administration i.e. there would be no management grant such as that paid under the original Dedication Scheme.

It was also proposed that local authorities should take a close interest in forestry planning—both State and Private. The Forestry Commission was instructed to consult with interests concerned with the aim of preparing proposals for a new scheme within the above framework for consideration by the Government.

Landowners, timber-merchants, industrialists and foresters were mostly critical of the Cost/Benefit Study and the conclusion in the Forest Policy paper that large-scale new planting of commercial woodlands could be justified only on the grounds that it created employment in areas of rural depopulation. There was widespread criticism too of the sudden cessation of grant-aid for private planting.

There was some criticism of the Forestry Commission which seemed to the outsider to have given in too easily to the anti-forestry lobby which was believed to exist in the Treasury and the Department of the Environment and its agencies such as the Countryside Commission.

Whatever the merits of the forestry case, Ministers were taken aback by the strength of the reaction displayed by landowners and forest workers alike.

A Forestry Commission press notice of 1973 says that 90 organisations and 60 individuals had sent their views on the 1972 Forestry Policy document [**3**] to the Commission. Obviously not all the views would have been pro-forestry, but clearly the majority were against the main proposals put forward by the three Forestry Ministers because the Government announced on 24th October 1973 [**5**] that they would re-introduce planting grants under a new Dedication Scheme covering woods of all sizes.

No mention was made about the creation of employment which was such a major point of the Ministers' own consultation paper. The need for private forestry schemes to conform to good land use criteria was now emphasised, together with recreational and environmental considerations. Ministers hoped—in vain as it turned out—that the rate of future planting would be similar to that which would have been obtained had there been no change in the arrangements.

4 The Possibility of Planning Controls

The great success of the planting programmes of the 1960s had some unfortunate repercussions in the 1970s. Environmentalists and planners had become more and more upset by the fact that the large-scale afforestation programmes were bringing great changes to the countryside, changes over which planners and environmentalists had no control. It was argued that, whilst the Forestry Commission would only plant up agricultural land with the agreement of the Agricultural Departments, the private sector, enjoying grants from the exchequer, were not even placed under this constraint.

The Countryside Commission for England and Wales, but not the Countryside Commission for Scotland, campaigned for forestry to be brought under planning control. Such a proposal could not, however, have the support of the Agricultural Departments whose officers recognised that if implemented it might lead to demands for planning control over agriculture, at least in the National Parks. Others felt that planning controls had not been successful in controlling undesirable urban developments and that they were costly to administer.

In the event—as has been noted—Ministers settled for a middle of the road compromise. Private forestry grants for new planting were to be given only where the appropriate Agriculture Department recommendations and Local Planning Authority views had been considered by the Forestry Commissioners.

Forestry Commission new planting schemes were also to be submitted to Local Planning Authorities for comment, the final decision being taken by the Commissioners.

The first public announcement that a change was being contemplated was made in 1972. It was only some seven years later that the private sector was able to see a clear road ahead.

5 Mechanical Development Committee

In 1971, the Home Grown Timber Advisory Committee's Technical Sub-Committee took over the work of the Mechanical Development Committee [**6**].

6 Forestry Training Council

Responsibility for forestry training in the private sector passed to the new Forestry Training Council set up under the auspices of the Forestry Commission during the year ended 31st March 1972 [**6**], following an announcement in August 1971 by the Secretary of State for Employment that forestry training should no longer come under the Agricultural, Horticultural and Forestry Industry Training Board.

The Council membership comprised members from the trade unions, private employers, forestry societies, professional organisations, the Training Boards and educational interests. The Council is voluntary and will not raise a levy but nevertheless is committed to the important task of raising the quantity and quality of training and supporting the associated further education in the forestry industry.

7 Forest Products Research Laboratory

In 1972 the Forest Products Research Laboratory, located at Princes Risborough in Buckinghamshire, was amalgamated with the Building Research Establishment. So ended with a stroke of the pen one of the world's most prestigious forests products research organisations that had served not only Britain during the war years 1939–1945, but its Empire and later many countries of the British Commonwealth which used its expertise in the development of their forest products economies when they became independent. Countries such as France with former colonial connections continued to maintain their own tropical timber research laboratories as part of the support for the exporters of sawmilling and other machinery and to continue to have an influence on the forest policies and forest politics of former colonies.

8 The Forestry Policy Statement of October 1973; New Dedication Scheme Promised

On 24th October 1973, The Minister of State, Ministry of Agriculture, Fisheries and Food made a statement [**5**] to the House of Commons on forestry policy:

"I am now able to give the Government's conclusions on the general framework (of future forestry policy). Consultations will continue on more detailed aspects.

Within our limited land surface and the increasing pressures upon it, the Government are concerned that forestry should form part of a really effective pattern of rural land use, in which it is harmonised to the best possible advantage with agriculture and the environment. Accordingly, the Forestry Commission, in administering grant aid to private forestry, will consult with the Agricultural Departments and local planning authorities on land use and amenity aspects of woodland owners' proposals. The Commissions's Regional Advisory Committees will be reconstituted to include representatives of agricultural and of local planning and amenity interests as well as members representing the forestry industry and will have the function of advising both on broad area strategy and on individual cases of doubt arising in the administration of grants.

There will thus be much improved arrangements for ensuring that proper weight is given to amenity considerations. Additionally, my Rt. Hon. Friends and I have asked the Chairman of the Forestry Commission specially to ensure that, in carrying out the Commission's own operations as well as in the exercise of their functions as Forestry Authority, every possible opportunity is taken to encourage and secure effective integration between agriculture and forestry.

Such is the importance which the Government and all those with whom they have consulted attach to land being properly used, and thus being well managed over the long term, that we propose to introduce a new Dedication Scheme more directly linked with these objectives. This single Scheme will succeed the previous Dedication, Approved Woodlands, and Small Woods Planting Grant Schemes and its terms will differ from those for woodlands that are already Dedicated. Thus in future, a Dedication covenant will normally terminate when approved felling takes place, but with provision for re-dedication. Owners entering the Scheme will receive an outright acreage payment as and when approved planting or re-planting takes place, in return for accepting a continuing obligation to manage the woodland concerned in accordance with plans of operations designed to secure sound forestry practice, effective integration with agriculture and environmental safeguards, together with such opportunities for recreation as may be appropriate. Departures from the plans of operations could involve recovery of grant wholly or in part. An appreciably higher rate of grant will be available for planting or re-planting where a significant proportion of hardwoods is involved.

Those having rights and obligations under the previous schemes will have the option of retaining them or of transferring to the new terms. It is the Government's intention that in any year the total of continuing payments under the old Schemes and of payments under the new arrangements should be broadly the same as would have obtained if the old Schemes had remained open to new entrants and planting had been at more or less the same rate.

My Rt. Hon. Friends and I hope that the remaining consultations will be concluded before the end of the year and further details, including rates of grant, will then be announced. In the meantime, the Forestry Commission will model its own planting and management policies upon the principles I have described; will give still further emphasis to providing recreational facilities; and will develop a planting and re-planting programme, on the lines and at the level set out in the consultative Document, and with special regard paid to the contribution which the Commission's operations can make towards stemming depopulation in rural areas."

Following the statement, the Forestry Commissioners began discussions on the detailed arrangements for the new Dedication Scheme with the forestry, agricultural and local authority organisations principally concerned.

9 Reconstructured Forestry Commission Accounts

9.1 Forestry Enterprise

In their 53rd Annual Report (1972–1973) [**5**], the Commissioners reported that: "The Forestry Commission has been in existence since 1919, a period of fifty three years during which 1 215 000 hectares of land have been purchased or leased, of which 749 000 hectares had been planted by March 1972. The historic acquisition cost of this land is £12.266 million while the accrued cost of growing the trees is £404.6 million; the total capital employed was £474 million.

In the accounts of the Forestry Commission it has been the practice to charge interest at the current Government lending rate irrespective of the earning power of the industry. £246 million or just over half of the above figure at March 1972 therefore represents accrued interest on money advanced over the whole of the fifty three year period. This figure of interest burden results from varying rates throughout the years and is equivalent to an average of 5% compound from 1919 to 1972.

A large proportion of Forestry Commission planting was undertaken for strategic and social reasons and, more recently, with increasing environmental impact. In many instances these plantings have been on relatively poor sites, often far removed from markets, and capable of yielding only a very low rate of return on the money invested.

It will be seen therefore that over a great part of the Forestry Commission's existence investment has been made for what were regarded as strategic and social reasons, without an expectation that a return at the current lending rate would be achieved. Most of the expenditure on the existing semi-mature forest estate has now been incurred and the strategic reserve concept has ceased to have the same relevance.

The Government has decided that it is therefore an appropriate time at which to restructure the Forestry Commission accounts, and to require in future that the Forestry Enterprise aim at a target rate of return on the funds entrusted to them. This rate has been set at 3% in real terms.

A rate of return of 3% in real terms is regarded as the maximum return that may be earned in Northern Europe on the best sites close to markets.

The historic cost of acquisition of land by the Commission up to March 1972 of course bears no relation to the value at which such land is currently changing hands. If the total Forestry Commission estate is divided into the historic cost the resulting average of £10 per hectare underlines the subsequent appreciation in the land value of the large estate which has been acquired for the nation.

Apart from its great recreation and amenity asset, however, the Forestry Commission estate is considered primarily in the financial context as a means of producing timber. It has therefore been felt most appropriate to arrive at a valuation of plantations determined solely by reference to an estimate of future expenditure and income discounted at a rate of 3%. The net discounted revenue so obtained comes to £128 million. The addition of £25 million for other assets produces a total of £153 million which represents the present economic value of the estate assuming that it will be managed for the production of timber.

The difference between 3% and the rate of interest charged to trading accounts of Government Departments will be recorded as a forestry subsidy each year and this will indicate the economic cost of putting these resources into forestry. In addition where less profitable activities are undertaken for reasons of Government policy, the Commission will be given additional subsidies each year to bridge the gap between 3% and the expected rate of return from any such activity. Those subsidies will be reviewed quinquenially when account will be taken of the effects of changes in the general level of prices.

The Forestry Commission estate will be revalued at five-yearly intervals and each valuation will be based on a physical inventory of the growing stock and the fixed assets. A comparison of successive valuations will show how the investment in forestry is developing and whether progress is in line with the achievement of the target rate of 3% in real terms. It will however be possible to see annually whether programmes are being achieved and to judge approximately how far the departures from the standards necessary to earn 3% in real terms are attributable to general movements in the level of prices and how far the effects of price increases are balancing those of cost increases.

9.2 Other Accounts

The opportunity has been taken to change the form of presentation of other accounts apart from those of the Forestry Enterprise. There is no longer a summary of the Forestry Fund because with the changing value of the currency a report of receipts and payments for the last 53 years is now not regarded as being of any significance.

Previously a Balance Sheet for the Forestry Authority was produced in order to give an arithmetical proof of the accumulated expenditure. This statement has been dispensed with and in its place the expenditure on the various sections making up the Forestry Authority is shown in detail and reconciled in total with the cash appropriated from the Grant-in-Aid of the Forestry Fund for 1972–1973."

10 The Forest Policy Statement of July 1974; Dedication Basis III Introduced October 1974

The next announcement on Forestry Policy had to await the outcome of a General Election and the settling in of a new (Labour) administration. On 5th July 1974 the Minister of Agriculture, Fisheries and Food made a statement in the House of Commons [**7**] which in effect accepted the proposals made by the previous (Conservative) Government and in addition to giving the level of grants to be paid, spelt out the details of consultative procedures.

The Statement read:

> "My Rt. Hon. Friends and I have decided to bring in new arrangements built upon the revised policy framework which was announced by the previous Administration on October 24th last and which included provision for the Forestry Commission's own operations. We attach great importance to the continued progress of forestry on these lines. In particular, we are introducing a new Dedication Scheme (Basis III) which has been devised after wide consultation with the interests concerned, including representatives of private forestry, planning, amenity, agricultural and other organisations. We aim to start on 1st October. An explanatory leaflet has already been prepared and copies are being placed in the Library of the House. Details of the arrangements are being notified to the European Commission.
>
> Owners entering the Scheme will accept a continuing obligation to manage their woodlands so as to secure sound forestry practice, good land use (including effective integration with agriculture), environmental benefits and provision of such opportunities for recreation (including provision for public access) as may be appropriate. In return, they will receive a basic grant of £45 per hectare (£18.21 per acre) payable after approved planting of an approved crop. For land approved for planting with hardwoods, and planted so as to establish a hardwood crop and give a predominantly hardwood ap-

pearance in the landscape over the greater part of the woodland's life, a supplementary grant of £125 per hectare (£50.59 per acre) will be payable in addition.

The Forestry Commission will consider for entry into the Scheme, both existing woodlands and bare land for planting, providing it normally extends to at last one hectare (approx. 2.5 acres). For entry into the Scheme of areas of ten or more hectares (approx. 25 acres) there will be a legal instrument of dedication under which, at the time of felling, either party will be able to exercise an option to end dedication. Similar, but less formal contracts are proposed for areas of less than ten hectares. In either case, a successor in title will customarily be invited to take up his predecessor's rights and obligations under a fresh agreement. Owners will be entitled to retain their existing rights and obligations by remaining under previous Schemes, or they can transfer into the new Scheme any area which meets its requirements.

In considering applications, the Commission will consult the Agricultural Departments, local planning authorities and other interests concerned, as appropriate, with the object of ensuring that land use, agricultural, amenity, recreation and nature conservation requirements are satisfied. They will also consult the planning and other authorities over environmental issues when considering owner's plans of operations. In handling individual grant applications and felling licensing, the Commission will endeavour, as at present, to secure the agreement of all parties concerned. However, in cases where differences of view remain, the Commission will invite the appropriate Regional Advisory Committee to assist them in the process of reconciling views; and if this further procedure fails to produce a settlement acceptable to all concerned, the Commission will refer the case to Ministers for decision. The Commission will also consult Regional Advisory Committees before commenting upon regional structure and local plans.

In the meantime, steps are being taken to reconstitute the Regional Advisory Committees to include representatives of agriculture, planning and environmental interests and the Forestry Commission will be announcing further details in due course.

The Government believe that the environmental safeguards built into this new Scheme will operate effectively, and in a manner which will command the confidence of Local Planning Authorities and, indeed, of all those having a special interest in the environment."

The most controversial point proved to be that concerning access. The undertaking required of an owner getting grants under Basis III was that he would be prepared to discuss with the Local Authority access arrangements for the public under the Countryside Act. In fact, such an undertaking could have no bearing on his legal rights and was a piece of window dressing to appease the "environ-

mental" lobby. In practice, in spite of their public posturing, Local Authorities have approached few owners on this score. Undoubtedly, a major deterrent is the fact that the Local Authority has to finance the creation and maintenance of any works necessary to allow public access and has to compensate an owner for any loss of revenue (e.g. from sporting) occasioned by the provision of public access.

A major disappointment for woodland owners was the absence of any grant for the maintenance of new plantations and the management of older woods. But there were other objections—often ill-defined—to the Scheme. The crisis of confidence created by the Forestry Ministers in June 1972 continued and fuel was added to the fire, when in August 1974 two Government papers on capital tax were published.

One proposed the abolition of the specially favourable "death duty" treatment accorded to woodlands (which became law in March 1975). The other announced the setting up of a Select Committee of the House of Commons to look into a tax on wealth. In December 1975, the Chancellor of the Exchequer announced that the Government would not introduce a Bill on Wealth Tax in the present session of Parliament, but of course the Government had rattled a sabre feared by most landowners.

One of the key features of the original post-war Dedication Scheme (covering Basis I and Basis II) was that owners dedicating their woodlands would bind their successors to remain in Dedication in perpetuity. In the event, for legal reasons, the Covenant the owner entered into was not to use the land for purposes other than forestry—a so-called negative covenant, but one binding in perpetuity.

Provision was made for possible release from Dedication in exceptional circumstances.

The Treasury never liked this long-term commitment which had been enthusiastically embraced by early post-war Chancellors of the Exchequer (e.g. Hugh Dalton). The Treasury were influential in changing the "in perpetuity" commitments of owner and government under Basis I and II to the choice for the successor in title to remain in Dedication Basis III or to terminate Dedication.

This fundamental change in policy aroused little controversy at the time—in part because the issue of the public's right to access to privately owned woodlands in receipt of government grant aid was very much to the fore and continued to be the subject of debate and discussion in the formulation of local authority regional and local plans in England and Wales and regional plans in Scotland.

10.1 Consultation with Local Planning Authorities

In the 1970s, the regional authorities which took the place of some of the old counties and those counties which remained in being were required to submit structure plans to central Government.

The Forestry Commission sought to have an input to the forestry content of these structure plans at the earliest possible stage. The prime responsibility for liaising with local authorities on the preparation of structure plans lay with Conservancies which in all cases consulted their Regional Advisory Committees as was required by government following the introduction of the Basis III Dedication Scheme.

The Regional Advisory Committees—re-constituted in 1974–1975 to include representatives of agriculture, planning and environmental interests as required by the Ministerial Statement of 5th July 1974—were in a position to offer the Commission advice on planning from a broader base of interests than the previous RACs with a membership made up almost exclusively of individuals from a forestry or timber trade background.

Naturally the forestry content of plans varied according to land use patterns, the local infrastructure, the degree of urbanisation and other macro-economic factors. But there were also influences on the plans which reflected local attitudes to forestry. For example, the favourable attitude to afforestation in Highland Scotland was in direct contrast to some areas in upland England where planners wanted to maintain the status quo in the use of unimproved land and where they persuaded the Ministers of Agriculture and the Secretary of State for Environment to announce that approval would not normally be given in these areas for any new planting which consisted primarily of conifers. In effect, this placed an embargo on the creation of new woodlands since most of the areas in question are suitable mainly for the growing of conifers on account of poor soils and exposure.

In Scotland there were suggestions by bodies interested in land use, that scientific surveys of the suitability of land for growing trees should be conducted and detailed forest land capability maps published. There were mixed views on the practicality and desirability of producing and publishing these detailed indicative forestry strategies.

10.2 Consequences in Scotland

Opponents of the idea—including some senior members of the Forestry Commission—whilst accepting that land-use surveys were a practical proposition, argued that to produce maps showing what forest crops the land was capable of producing somehow infringed on the owners rights to use his land as he wished and might in some cases affect the value placed on an estate for tax purposes e.g. for inheritance tax. Others felt that such scientifically based land capability surveys would (a) be beneficial additions to the criteria on which local authority planners might make judgements on the suitability of land for planting (as opposed for example to sheep grazing) and that (b) they would help to prevent local authority judgements being made solely on amenity criteria (such as "no tree planting to be allowed within X miles of public roads so as to provide views unhindered or unspoilt by trees").

Note: In 1989 the Forestry Commission changed its attitude and commissioned the Macaulay Institute in Aberdeen to identify and describe the land area of Scotland according to its suitability for the growth of trees of different species. Five maps (1:250 000) and their accompanying booklets are now available from the Forestry Commission or the Macaulay Land Use Research Institute, Craigiebuckler, Aberdeen.

Under the aegis of the (Scottish) Standing Committee on Rural Land Use in 1975 the Forestry Commission made a study of the effect of factors influencing tree growth in Scotland by geographic regions so as to assess the total area that might be afforested in addition to the area already under trees.

Note: The (Scottish) Standing Committee on Rural Land Use, was formed in consequence of the Governments White Paper, Cmnd 5428, Land Resource Use in Scotland—The Government's Observations on the Report of the Select Committee on Scottish Affairs [**8**]. It was chaired by the Department of Agriculture and Fisheries for Scotland (DAFS) and included representatives from DAFS, the Scottish Development Department, the Scottish Economic Planning Department, the Countryside Commission for Scotland, the Forestry Commission and the Nature Conservancy Council.

Topography, soils, climate and the areas under and likely to remain under agriculture were taken into account. The main conclusion of the Report which was issued by the Standing Committee to all Regional Authorities [**9**] in Scotland to assist them in the formulation of their plans, were summarised thus:

"If there are physical limitations to the extent to which forestry can be practised in the uplands, equally there are economic limitations to the extent to which it can be practised on the more fertile sites at lower elevations. The latter areas are primarily agricultural, but also contain the major centres of population together with their associated industries and communications. Nevertheless, this low ground already holds substantial areas of woodland, much of it in the form of policy woods or shelterbelts, although there are also plantations grown primarily for timber production. A marked increase in woodland area in these predominantly agricultural and urban situations seems unlikely, but the existing woodlands, as well as the single trees in the hedgerows and gardens, contribute far more to the character of the landscape than their area and number might suggest.

With the low ground largely excluded from forestry for agricultural and urban use and the topmost ground excluded because of exposure, low temperature and shallow soils, forestry becomes essentially an industry of foothills and the uplands. Land for further expansion is therefore limited largely by physical factors. An estimate has been made of the area in each Region which is considered capable of growing timber crops. These are shown in Table 5.2 and it will be seen that in some instances, a considerable proportion of the area is already under woodland.

Because a considerable amount of the existing woodland area occurs on low ground, the net area available for commercial extension is consequently underestimated, but it would appear that there are about 1.7 million hectares of land,

Table 5.2 Estimated area of land potentially suitable for afforestation in Scotland in 1976

Region	% area considered suitable	Gross area suitable ($\times 10^3$ ha)	Already woodland ($\times 10^3$ ha)	Net area suitable ($\times 10^3$ ha)
Dumfries & Galloway	47	300	100	200
Borders	42	200	50	150
Lothian	31	55	15	40
Strathclyde[a]	40	560	160	400
Highland[b]	27	705	205	500
Central	43	115	35	80
Fife	15	20	15	5
Tayside	39	300	80	220
Grampian	22	190	130	60
Western Isles	2	5	—	5
TOTAL	—	2450	790	1660

[a] Excludes all islands except Mull and Arran
[b] Excludes all islands except Skye

currently under some other form of land use, which could be planted with trees to produce economic crops of timber.

If fully implemented, a major land use change of this nature would undoubtedly affect the upland agricultural industry—in particular the hill sheep and, to a much lesser extent, the hill cattle sectors—to a considerable degree. However, as much of the potentially plantable land is of relatively poor quality and so is currently carrying stock at fairly low density, and as the change would be gradual rather than sudden, the overall effect on agricultural production from the uplands is likely to be much less than the area involved might initially lead one to suppose."

Over the years broad indicative forestry land use studies have come to be accepted and forestry has undoubtedly benefited from the discussions which have taken place between foresters and local authorities.

In time the early pioneering work of the Forestry Commission and the Scottish Development Department in encouraging Regional Councils in Scotland to include some indication—however general—of the possibilities for new planting, bore fruit. For example, in their 70th Annual Report [**10**] the Commission reported as follows:

"In recognition of the importance to Scotland of the development of forestry, the Secretary of State announced in February 1989 that he would be encouraging Regional Councils to prepare indicative forestry strategies as part of their structure plans."

11 Forestry Safety Council

The Council was set up on 1st April 1974 [**11**]. Its membership included persons nominated by the main bodies representing the growers and timber merchants in Great Britain; representatives of workers in the forestry industry; the Forestry Commission and the Forestry Training Council.

The purpose of the Forestry Safety Council is, in relation to all aspects of safety in the forestry industry generally to:

- Promote safety
- Make representations to the Safety Authority and be consulted by them on legislation, regulations and other relevant matters.
- Produce and maintain codes of safe working practice.
- Co-ordinate publicity and collection of statistics on health and safety.
- Encourage research into health and safety.
- Liaise with the Forestry Training Council and other appropriate bodies.
- Report annually to the Forestry Ministers through the Forestry Commissioners and the Home Grown Timber Advisory Committee.

The Commission had actively participated in the work on health and safety of forest workers at the international level since the early 1950s, through the very successful Joint ECE/FAO/ILO Committee on Forest Working Techniques and Training of Forest Workers of which it was a founder member.

The membership of the Joint Committee included the Nordic countries, Europe, North America and the former Soviet Union.

12 The Commission's New Objectives

In their 54th Annual Report (1973–1974) [**7**] the Commissioners announced the redefinition of the objectives of the Forestry Authority and of the Forest Enterprise, based on work which paralleled that on the new grant-aid arrangements for private forestry. The Commissiones' statement of their objectives is given below:

"The activities undertaken by the Forestry Authority follow from the Commission's general duty of promoting the interests of forestry which includes the dissemination of forestry knowledge, the promotion of the supply of timber, the conduct of research, the administration of schemes of grant, or other aid for private forestry and the protection of woodlands through planting, felling and plant health controls; in all these functions, attention is paid to the desirability of conserving the natural beauty and amenity of the countryside.

In discharging these functions, the specific objectives of the Forestry Authority are:

(a) to advance knowledge and understanding of forestry and trees in the countryside;

(b) to develop and ensure the best use of the country's timber resources and promote efficiency and development in the home timber industry;
(c) to undertake research relevant to the needs of forestry;
(d) to combat forest and tree pests and diseases and to initiate Plant Health Orders when appropriate;
(e) to advise and to assist with safety and training in forestry;
(f) to administer controls and schemes for assisting private woodland owners and by so doing, encourage the practice of sound forestry, secure good land use and—where relevant—effective integration with agriculture and ensure the use of forest management systems and practices which safeguard the environment.

The main general objective of the Forest Enterprise remains that of producing wood as economically as possible. However, in pursuit of this aim, as in the case of the Forestry Authority, attention is paid to amenity, and the Commission is required to preserve and enhance the landscape and to develop the potential of its properties for nature conservation and recreation.

The objectives of the Forest Enterprise are:

(a) to develop forestry and increase the production of wood for existing industries, or industries yet to be established, by the extension and improvement of the forest estate;
(b) to protect and enhance the environment;
(c) to provide recreational facilities;
(d) to stimulate and support the local economy in areas of depopulation by the development of forests, including new plantations and of wood-using industry;
(e) in pursuit of these objectives and in the extension of the forest estate, to further the integration of forestry and agriculture and to manage the estate as profitably as possible."

The Commission went on to expand on some of the broader considerations which lay behind their formulation of these policy objectives.

The following paragraphs summarise the key statements made by the Commission.

12.1 Wood Production

The production of timber continues to be the prime objective of the Commission and represents by far its major source of revenue.

12.2 The Landscape

No one sector of the community nor single generation has a monopoly interest in the forest and its management. The forester is a trustee for the community at large with a duty to pass on our inheritance from the past, avoiding change for its own sake and shaping the inevitable cycle of growth, maturity and replacement in a manner which blends the best of tradition, currently accepted standards and options for the future.

Forestry Commission policy is to perpetuate the broadleaved character of the British scene where this already exists. There is a role for conifers even in areas such as the southern half of England where many of the existing pine, spruce and fir woodlands are highly prized by both residents and visitors.

In upland areas where soil and situation often make the choice of conifers inevitable, every opportunity will be taken to ensure effective landscaping.

The use of broadleaves and landscaping of conifer plantations inevitably reduce the financial return at rates varying according to local circumstances, but some reduction is accepted by the Commission and the Government Departments concerned.

12.3 Conservation

It is the Commission's aim to improve its forests and wildlife habitats and to integrate nature conservation and wood production in a pattern of good land use.

12.4 Recreation

Recreation benefits can be achieved with minimum direct loss in the production of timber, especially those associated with the enjoyment of the natural environment of woodland. Priority will be given to the value of the forest to the day visitor.

12.5 Areas of Depopulation

Labour productivity has risen rapidly but the creation of new wood using industries will call for increased local harvesting and transporting of timber.

12.6 The Integration of Forestry and Agriculture

Note: The "integration" of forestry with agriculture was a concept that became fashionable in some Government Departments at this time, although many officials were at a loss to know precisely what was to be involved in practice. The Forestry Commissioner's own interpretation of the Government's integration policy aims, shows a certain lack of enthusiasm, as may be judged from the verbatim extract from the 54th Annual Report [**7**]:

"Although this objective is well understood and widely accepted in a general sense, has been practised on progressive estates for generations and been the subject of many case studies which have clearly brought out the nature of the benefits, it has not yet proved possible to establish and quantify definitions and criteria for the assessment of an integrated project which are universally applicable at the planning stage. In the short term, the contribution to improvement of agriculture by the fencing and roading required by forestry may be the most important factor; in the medium term, this latter may be the shelter provided by the growing trees; and in the long term the contribution to the farm's income from harvesting the mature trees; and, in many cases, from a more generally productive use of the land by more intensive agricultural use of areas remaining unplanted. Individual proposals will be considered by Conservators in close consultation with farming organisations and the Agricultural Departments."

Writing some 20 years later, the benefits of integrating forestry with traditional livestock (mainly sheep) raising have been expressed by the (Scottish) Crofters Commission and the Scottish Crofters Union following on the removal of legal restraints (via the 1991 Crofters Forestry (Scotland) Act) and the provision of a more realistic level of grants to help crofters establish and maintain new woodlands on crofting land (see Chapter 9, including References).

These benefits are briefly summarised here:

Why Plant Trees?

> Woodlands are an important part of the natural landscape in many parts of the Highlands and Islands and well designed planting can enhance the wildlife value of an area. As well as providing shelter for livestock, buildings or crops, trees can hide unattractive features or provide scenic interest. They can also provide useful wood in the future and may use small, awkwardly shaped or steeply sloping areas of land and prevent soil erosion. Trees can make a positive contribution to the local economy through the work involved in tree planting or future woodland management, the impact on tourism in the area and the eventual timber products.

12.7 Profitability

The Commission's financial objectives will be directed to the targets set by the reconstruction of the accounts (paragraphs 14–23 in the 53rd CAR [**5**]). Costs in the field have been held by new management methods and technical improvements and efforts to this end will continue, as will those towards the containment of overheads that are within the Commission's control. In the short term the rise in timber prices may make it easier to attain the financial target set, but this results from the world situation and cannot be relied on indefinitely. The objective of managing the forest estate as profitably as possible may be subject to certain calls on resources: in the longer term a need for the Commission to encourage by appropriate means the development of wood-using industry may constitute such a call; the other social objectives set out by Ministers, and referred to above, so far as they may not justify subsidies, may form another.

13 The 1976 Review of Grants and Taxation of Woodlands: Inter-Departmental Review Group

The Finance Act 1975 repealed the greater part of the former estate duty provisions for forestry and replaced them with capital transfer tax [**12**]. Private planting dropped dramatically in 1976 to just over half the 1975 figure (Table 5.1) and by July 1976 the Government was sufficiently impressed by representations made to it on behalf of the private forestry sector to promise to review the whole forestry taxation and grant-aid field.

Without doubt the Government—in coming to this decision—was influenced by the support for private forestry that had come from the ordinary forest workers who joined in a well-conducted and hugely effective pro forestry demonstration at the House of Commons. It was said that Treasury and other Ministers were upset by the bad briefing they had been given by their officials who had painted private forestry as a rich man's hobby and a tax avoidance device with which the ordinary man in the countryside had little sympathy. Now forest workers had turned out in their thousands to support their industry.

The highly professional forestry management companies that had channelled so much wealth into forestry in response to the governments post-war forestry policy initiatives (notably the Dedication Basis II Scheme) were willing and able to articulate the case for continuing government support for the expansion of forestry in a more aggressive way than had traditional landowners.

Once it had decided to carry out the review, the Government moved quickly to set up an Inter-Departmental Review Group under Treasury chairmanship with the following terms of reference:

> "To review in the light of changes since 1972 in capital taxation and in the grant arrangements through the Dedication scheme the taxation and grant arrangements for private forestry taking account of economic, fiscal and environmental considerations; and to report by 31st December, 1976" [**13**].

The Departments represented in addition to the Treasury were Inland Revenue, Ministry of Agriculture, Scottish Office, Welsh Office, Department of the Environment, the Forestry Commission and the Department of Agriculture, Northern Ireland.

Representations were made to the Interdepartment Review Group by a large number of organisations and individuals, some of whom were invited to give oral evidence. The written and oral evidence was well argued and constructive and resulted in the Group suggesting measures which met some of the objections to the new tax regime for forestry put forward by the private sector.

It soon became clear to those giving oral evidence to the Group, that the Group had achieved a far greater degree of understanding of the factors which motivated private individuals to engage in the very long-term business of forestry than had previously been evident in official circles. Many felt that at last the forestry case had got through to the Treasury. In their 1977–1978 Annual Report [**14**], the Commissioners quote the Forestry Committee of Great Britain (representing private woodland owners) as being grateful for further adjustments to the taxation of woodlands rules "completing measures flowing from the work of the Inter-Departmental Review Group". In the next section the Government's positive response to the Review Groups recommendations is summarised.

14 1977—Increased Grants: Small Woods Scheme Re-introduced

On 30th March 1977 the Minister of State, Treasury announced in a written answer to a Parliamentary Question [**13**], that there would be very substantial increases in the level of planting grants for Dedicated Woodlands; a management grant would be introduced for the new (Basis III) Dedication Scheme for which only planting grants had been payable under the 1973 rules. A small woods planting grant would be re-introduced for woods up to 10 hectares (25 acres) in extent; this would save owners of such woods from having to Dedicate them with all the legal complications, costs and delays. To encourage the planting of broadleaves, grants were to continue to be higher than those for conifers under Basis III Schemes.

Such differentiation was too costly to administer for the Small Woods Scheme, but it was made clear that the Forestry Commission were to encourage broadleaved planting under this scheme and could refuse a grant in situations where an owner could and should plant broadleaves but would not.

On 31st March, the following day, the Parliamentary Secretary to the Minister of Agriculture, Fisheries and Food said:

"Timber production to feed our wood-processing industries continues to be recognised as the ultimate goal for woodlands managed under the (Basis III Dedication) Scheme, as under the previous Basis I and II Schemes; but in pursuing this important end objective woodland owners are called upon to temper their management practices so that private forestry develops in full harmony with agricultural and amenity requirements."

The Forestry Commissioners commented that "This re-emphasis of forestry's principle objective will clearly be of encouragement to the private sector of the industry."

New planting in private woodlands, perhaps the best single indicator of confidence among private forest owners, had fallen to 7200 hectares in 1977 and to 6300 hectares in 1978, compared with 20 000 hectares in 1972. Two to three years are needed to prepare land for planting, plant it, have it inspected by the Commission and passed for payment; and any substantial increase in demand for plants requires increased sowing in the nursery followed by a year or two's growth in seedbeds or transplant lines. Thus new planting recorded by the Commission in 1979 at 7900 hectares could be said to reflect the decisions of owners taken in 1977 and to be the earliest indication of any real revival of confidence following the 1977 review of private forestry. By 1980 new planting had risen to 8300 hectares.

What had been gained in the 5 years from 1972 when all grant aid to private forestry (other than legally binding commitment of then existing Dedication Schemes) had ceased and the unsatisfactory Basis III Dedication Scheme had been introduced in October 1974? At the expense of the creation of a great deal of uncertainty in the private sector over several years, the anti-forestry lobby evident in certain Government political and official circles, had been defeated. Ministers of the two main parties—Conservative and Labour—had learnt that the views of those of their advisers who were anti-forestry were off-set by a pro-forestry lobby much stronger and much more widely based than they had been led to believe. On the other hand, there were positive gains for those moderates in Government service who thought that commercial forestry should be subject to some form of control, so that in the event of a planting proposal being contrary to good land use, aesthetically unacceptable in terms of the local landscape, or harmful to local nature conservation, grants could be refused by the Commission acting as Forest Authority. In the process of confrontation, much was learnt by foresters and about foresters and forestry. A more statesmanlike approach might not have had that effect and might have simply put off the day when forestry was subject to critical scrutiny.

The Forestry Commission could take comfort from the fact than irrespective of the somewhat gloomy findings of the Cost/Benefit Study of 1972 and of their own worries about the cost of administering grant aid to private forestry, they had received a definite mandate to continue to pursue a positive policy of expansion of the forest estate whether by their own direct afforestation schemes or indirectly through the grant aiding of private forestry.

15 Dutch Elm Disease

Dutch elm disease—which kills elms—is caused by two related species of fungi (in the genus Ophiostoma) which are spread by various elm bark beetles [**15**]. It is met with in many countries and was recorded in Britain in the 1920s. But by the 1940s the epidemic, which had peaked in the 1930s, had declined to the extent that it was regarded as an endemic problem of no great importance.

In the late 1960s a new epidemic developed. It was caused by the introduction to Britain on diseased logs imported from Canada of a more virulent form of the Dutch elm disease fungus than that which had been responsible for the 1930s epidemic. Within 5 years some 30% of the elm population of southern and central England was dead and dying and by 1977 it was reported by the Commission that despite strenuous efforts here and in North America, no cure had been found.

In their 58th CAR (1977–1978) [**14**] the Commission reported the spread of the disease into hitherto lightly infected and unaffected areas in Wales, the North of England and Scotland, notably the Border counties. By 1980, it is estimated, over 20 million elms had been killed.

Throughout the 1970s various restrictions on the movements of elm timber with the bark attached were introduced and enforced (e.g. 8 offenders were successfully prosecuted in the year 1978/79 and 13 more cases were pending). By this and other local measures it was possible to delay the spread of the disease into lightly affected areas. As the disease continued to spread over the next decade, 1980–1990, the number of local authorities financially willing and physically able to control the spread of the disease through sanitation fellings, root severance and other measures in lightly infected areas diminished. The range of the disease has now extended northwards in Scotland, although it is not known to be present along the Moray Firth and north of the Great Glen.

There are several species of elm in Britain, most of which sucker freely and the death of the original elm is often followed by vigorous regrowth which in turn becomes infected by the fungus which can remain viable in the old root system for many years.

In the uplands and the north the native wych elm occurs. It does not reproduce by suckering but from seeds. It is therefore less prone to infection via root transmission than suckering species. Added to this the fact that its bark is less attractive to the elm bark beetle (which in turn is adversely affected by lower temperatures) helps explain better survival rates in certain areas.

Prior to the ravages of Dutch elm disease, the wood of elm was in demand for a variety of uses of which the most specialised was for coffin boards where its availability in good widths was an important factor. Its interlocking grain and relatively good durability under water meant that it was used for rubbing timbers on dock-sides; for canal gates and their associated woodwork; and for river and sea defences. Its attractive grain gave it a niche in the furniture market.

None of these markets could increase their consumption significantly when supplies of dead and dying elm increased overnight in the 1970s (the quality of the timber is quite unaffected by Dutch elm disease).

The Forestry Commission formed an Elm Marketing Group which comprised specialised hardwood timber merchants and Commission specialists in marketing in an effort to maximise the use of elm in its traditional markets and promote new outlets. It carried out a survey of sawmills in England and Wales capable of handling elm and built up a register of potential users and suppliers of round and sawn elm. The Group did useful work, including helping in the export of regular and substantial quantities of elm to French furniture specialists making solid elm cafe and restaurant furniture of traditional design.

Research in various countries into means of controlling the bark beetles and the fungus and into breeding elms resistant to beetles and fungus have so far been unsuccessful. Genetic engineering to produce disease resistant English elm is a recent British initiative.

Figure 5.1. Front cover of Forestry Commission Research Publication [15]

The Disease Cycle—Forestry Commission Research Information Note 252 [15]

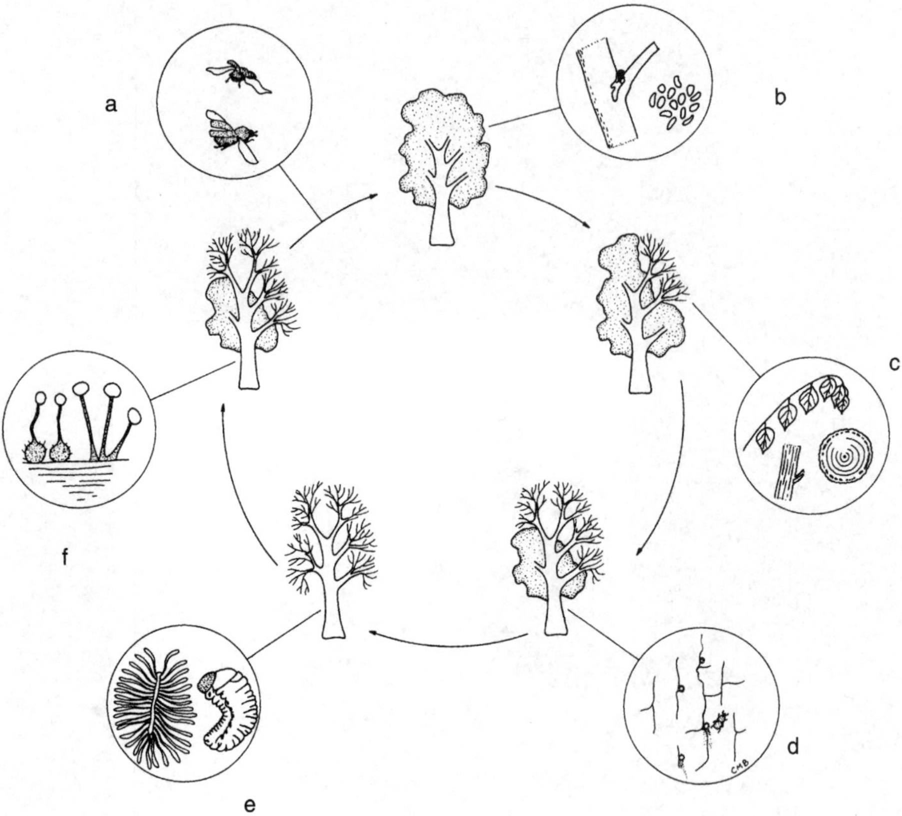

Figure 5.2. Life cycle of fungus and beetle:

(a) Adult beetles emerge in spring and summer from the bark of dead and dying elms carrying spores of the causal fungus.
(b) They feed in the twig crotches of healthy elms and introduce fungal spores into the wood.
(c) As a result the infected parts wilt and diseased twigs show characteristic dark streaks or spots.
(d) Trees weakened by disease become breeding sites for beetles.
(e) The larvae develop galleries within the bark.
(f) The fungus fruits in the pupal chambers.

Figure 5.3. Root transmission through a hedgerow of English elm

16 The Countryside Review Committee

In 1974 the Department of the Environment set up, with the approval of the then Secretary of State, a Countryside Review Committee, with the following terms of reference:

> 'To review, in relation to land outside urban areas, the state of the countryside and the pressures upon it;
>
> To examine the effect of existing policies for, or having an impact upon, the countryside and the extent to which they are adequate to contain or modify or accept the pressures;
>
> Given the existence of other major policy objectives, including the maintenance of agricultural production, to consider whether changes of policy or of practice are necessary to reconcile these objectives where they conflict with the conservation of the countryside, the enhancement of its natural beauty, and its enjoyment by the public;
>
> And to make Recommendations'.

In 1979 the Committee noted [**16**] that: chaired and serviced by the Department of the Environment, the Committee includes among its members the relevant planning, conservation, recreation and regional interests of that Department. The Ministry of Agriculture, Fisheries and Food, the Department of Transport, and the Welsh Office are also represented. The Committee's remit does not extend to Scotland, but the Scottish Development Department has provided observers.

Representatives from other Departments have attended meetings as necessary, and have been consulted by correspondence. Officers from the main public Agencies concerned—the Countryside Commission, Nature Conservancy Council, Development Commission, Forestry Commission and Sports Council—have served on the Committee.

Topic paper 4 Conservation and the Countryside Heritage, A Discussion Paper, was published in 1979 [**16**].

From the forestry point of view, its most important conclusion was (as stated in the summary)

> "In the wider context there have been pressures for planning control over agriculture and forestry; however the Countryside Review Committee does not believe that such controls would be justified or would serve a useful purpose."

Looking ahead, we find that within 5 years the Countryside Commission recommended that forestry schemes of more than 50 hectares in the uplands should be subject to planning control—a recommendation that was not accepted by the Government [**17**].

And in 1986 the Government rejected the Nature Conservancy Council's recommendation that planting should be banned in SSSIs, pointing out that there was already a strong conservation commitment by the Forestry Commissioners in managing their own land and giving grant aid to the private sector. In like manner the NCC's criticism of large-scale afforestation was rejected on the basis that there was scope in Britain for such schemes provided a balance was reached with other land use interests [**17**].

References

1. 51st CAR
2. 52nd CAR
3. *Forestry Policy.* June 1972 HMSO
4. H M Treasury. *Forestry in Great Britain. An Inter-departmental Cost/Benefit Study.* 1972 HMSO
5. 53rd CAR
6. 52nd CAR (page 35, page 45)
7. 54th CAR
8. Land Resource Use in Scotland—The Government's Observations on the Report of the Select Committee on Scottish Affairs—Cmnd.5428. HMSO
9. Forestry Commission. *The Place of Forestry in Scotland Research and Development Paper 113.* Published February 1976
10. 70th CAR
11. 55th CAR (page 44)
12. 56th CAR
13. 57th CAR
14. 58th CAR
15. Forestry Commission. *Dutch Elm Disease in Britain.* Research Information Note 252. Alice Holt Research Station, Wreccelesham, Farnham, Surrey
16. Countryside Review Committee. *Conservation and the Countryside Heritage. A Discussion Paper.* Topic Paper No. 4. London HMSO 1979
17. *The Forestry Commission. The First 75 Years.* 1994

Chapter 6

Forest-based Recreation and Public Access

The policy of the Forestry Commissioners has been to give the general public access on foot to its forests, unless, as is the case in a minority of woodlands, the land is leasehold with legal constraints on public access. Even here the Commission has over the years bought out leases so as to remove such restrictions.

As a means of actively encouraging people to enter on its land, the Commission conceived the idea of designating National Forest Parks, where the public was offered:

- free and unlimited public access to large areas where people might enjoy walking over wide stretches of countryside.

The first of its forest parks—the Argyll National Forest Park—was designated in 1935.

In this context one has to remember that in acquiring land suitable for planting the Commission might have to buy with it large areas of open land unsuitable for afforestation but well suited to walking. Although it was not then envisaged that such a wide range of activities would be made available as is now the case, there was a strong commitment to a forward looking recreational policy that anticipated a growing demand for countryside leisure pursuits by a largely urban population; and it was recognised that it is the ability of forests (as opposed to open land) to provide informal recreation that gives them a key role in providing for the needs of large numbers of people. The interdepartmental Cost/Benefit Study of 1972 [**1**] said:

"Turning now to the less specialised forms of recreation, we find a very different picture (from specialised pursuits such as shooting and fishing enjoyed by relatively few people)—large numbers of visitors, most of whom produce no revenue for the Commission. There was a large measure of agreement between all the bodies consulted on the potential role of the Commission in this field. Trees provide shelter from wind and rain, a very useful insurance in all seasons of the year. More importantly, trees add very significantly to what planners term "the recreational capacity" of an area. Forests are about the only effective natural way of absorbing motor vehicles into the landscape. They provide a valuable degree of privacy to those for whom the country implies an escape from other people. But open spaces and glades cater just as well for the more congenial."

Here the study lists five areas in southern England near large centres of population that attract thousands of visitors at weekends whose capacity is probably exceeded only by the more popular seaside resorts.

Between 1935 and 1955 six more National Forest Parks[1] were designated—Snowdonia (1937), Dean and Wye (1938), Galloway (1943), Glenmore (1948), Queen Elizabeth (1953) and in 1955 Border. Since 1924, when the Commission took over its management, the New Forest has been treated as if it were a National Forest Park [**2**].

As the Commission approached its 50th anniversary in 1969 [**3–5**] it had, in response to demand, already begun to widen its range of recreational facilities, encouraged by the fact that the 1967 (Scotland) [**6, 7**] and 1968 (England and Wales) [**8, 9**] Countryside Acts strengthened the Commissioners specific powers to provide facilities for recreation. The Commissioners, in their Annual Report for the period ending March 1969 [**8, 9**] said:

> "The Commissioners continue to lay the strongest possible emphasis on the conservation of amenity and the provision of recreational facilities, in the belief that, so far from amenity and recreation being incompatible with commercial forestry, they are now permanent and integral elements in its practice."

In the 1960s the European Forestry Commission (EFC) of the Food and Agriculture Organisation (FAO) initiated a series of debates on the extent to which a forest realistically could fulfil more than one function—such as production of wood, amenity, nature conservation, recreation, tourist control (by providing "honey pots" to lessen pressures elsewhere) sporting and amelioration of climate. The majority view was that most forests could fulfil a variety of functions without any one function suffering unduly, but for each forest an order of priorities had to be established between the various functions or potential functions that the forest could offer. Generally the Commissioners did not lay down a list of priorities for individual forests and went only so far as to assert that recreation, amenity and timber production were generally compatible activities.

They have been shown to be right in that the conflicts that have occurred since this 1969 statement have been mainly between different forms of recreation, not between timber production, recreation and, with few exceptions, amenity. This has meant that in some instances, such activities as horse-riding, cycling and walking have had to be allocated separate routes through the forest.

---

[1]The word National was removed from the titles of the Forest Parks when National Parks as such were set up in the early 1950s.

Motor rallying falls into a different category from other activities and motor rallies take place (for short periods) only on pre-determined forest roads and a forest may see a rally only once in a period of years, or never.

In 1970 the Commission set up a Conservation and Recreation Branch at its Headquarters [**3, 4**]. Its initial task was to formulate a recreational policy and develop:

> "The unique recreational features of forests for large numbers of visitors, whilst both safeguarding the primary objects of timber production and ensuring that the recreational developments do not conflict with the conservation of the forest environment."

A Recreational Planning Officer was appointed to plan the implementation of recreational facilities, after consultation with local planning and other authorities.

Three sorts of facilities were formally identified. First, of course, were the long-established day facilities for those seeking fresh air and exercise. These were identified as car parks, picnic spots, way-marked forest trails and the extensive network of forest roads (for use by pedestrians). Second were 'special activities', such as orienteering, pony trekking, forest drives, wildlife study and fishing. Third were caravan and camp sites, forest cabins and holiday homes, which had to earn a rate of return of 10% on capital employed. Particular attention was devoted to building forest cabins.

Not all Commissioners and not all politicians supported the extension of the Commission's over-night facilities to include forest cabins and holiday cottages, but they were in a minority. Throughout the country, the potential for building new holiday complexes and letting surplus houses was assessed using a 10% test discount rate.

Whereas in the immediate post-war years every field forester and area manager was under pressure to increase his planting programme, so now was there pressure to identify caravan and camping sites, car parks, picnic areas and other recreational facilities.

At this time too, the Commission began seeking for a corporate image in the signposting of its individual forests and the facilities they had to offer.

An important criterion in the allocation of funds to new recreational facilities—whether residential or other—was that priority would be given to schemes nearer to large centres of population. This marked a departure from the earliest priorities of giving access to the remoter, wilder, generally upland areas.

Looking ahead, in the 1980s, the Conservative government stopped the Commission's programme of building its highly successful forest cabin holiday centres for political reasons. Those that had been built and the many caravan and camp sites remained in being and are much sought after. The numbers in being in 1994 and 2000 are noted at the end of this chapter.

What of the private sector? There are many miles of public rights of way in private woodlands, but apart from their obligations to recognise these, the question of owners allowing the public access to their woods as a quid pro quo for their receiving Government grant aid or tax relief was an issue not raised until the 1960s. There were many reasons for this, not least that the priority was to encourage owners to restock their woodlands following the fellings of the first and second world wars. It was also realised that many private woodlands did not lend themselves to public access. For example, many were small in area, often intimately bound up with farmland and the owners residence and not uncommonly having an important sporting role on the estate or farm.

On some of the larger estates, notably in Scotland, by tradition public access to parts of the estate woodlands was encouraged or tolerated; some have provided picnic areas and walks (sometimes in conjunction with local authorities).

In the early 1970s, pressure from the Department of the Environment eventually persuaded the Government to make it a condition of their receiving grant aid (and at that time tax relief) that owners entering new schemes would at least be willing to discuss with local authorities the question of public access to their woodlands coming within the schemes. The Government announcement of the new Dedication (Basis III) Scheme on 5th July 1974 confirmed this requirement [**10**].

The often highly politicised debate still continues with many landowners feeling that the availability of so many Forestry Commission recreational opportunities is a major factor in taking the pressure off the private sector.

There are no official statistics to indicate the extent to which the public have access to private woodlands, but in 1990 in response to a United Nations inquiry [**11**], the Forestry Commission as Forestry Authority gave an assessment of the relative importance of the various functions of the State (FC) and private forests of the UK. In public forests (mainly Forestry Commission of Great Britain and Forestry Department of Northern Ireland) recreation had high importance on half the total area of 923 000 hectares surveyed and medium and low importance respectively on the two remaining quarters.

By way of contrast, in private woodlands recreation was of high importance on only 15% of the total area of 1.28 million hectares; of medium importance on 10%; and low importance on 75%. These figures do not indicate to what extent there is free public access to or free recreational facilities in private woodlands, nor to what extent private owners provide recreational facilities against payment, but they do indicate that by and large private woodlands are still private woodlands in the UK.

Some idea of the scale of Forestry Commission recreational facilities is given by the figures below. In 1994 there were: 741 forest walks and nature trails and 629 picnic places, all open to the public without charge [**12**]. Some 29 visitor centres giving information on local and regional Forestry Commission facilities, history and current activities were supported financially by sales of forestry and general countryside literature and souvenirs.

Recreational facilities for which a charge is made and which are expected to yield a return on the capital invested in them included camping and caravan sites (29 plus 33 for youth camping; 174 holiday cabins and houses); and 13 forest drives where the modest charges cover the road maintenance.

Charges are also made for orienteering and wayfaring (57 courses); and the 33 wildlife hides.

Charges, if any, are set locally for the 143 cycle trails and 64 horse trails.

There were also 21 arboreta and forest gardens, including some of the best known such as Bedgebury and Westonbirt, managed by Forest Enterprise and with admission fees to each arboretum set locally to reflect its size, scope and running costs.

By the turn of the Century there were some changes in the number of recreational facilities—due in part to sales of Forestry Commission land and some in response to changes in public demand (e.g. for cycle trails). The numbers of recreational facilities as at 31st March 2000 are given below [**13, 14**].

Forest Walks	703
Cycle Trails	392
Horse Riding Routes	114
Forest Drives	10
Picnic Sites	507
Orienteering Courses	82
Visitor Centres	31
Wildlife Hides	27
Caravan/Camping Sites	25

Self-catering holiday log cabins were unchanged at 166 but the few holiday houses had by 2000 been sold off. Holiday cabins continued to achieve occupancy levels "well above self-catering norms" [**14**]. Discussion of the use of woodlands for recreation centres around the questions of wider free access for the general public which private landowners are generally unwilling[2] or unable to offer; and of informal recreation [**15**] for which they may well make a charge.

Where free access—as on Forestry Commission land—is enjoyed by the public, the benefits that are created have a value, but as there is no market where these social benefits can be traded, evaluating them in monetary terms is not easy [**16**] but must be attempted. The recreational value of the Commission's forest estate is discussed in the Interdepartmental Study of 1972 on Forestry in Great Britain [**1**], where it is argued that the recreational value of the Commission's estate could be as high in some parts of the country as its value as a source of timber.

[2] On many estates, especially in southern England, the combination of woodlands and agricultural land create ideal conditions for shooting pheasant on a commercial scale. The revenues from these shoots can be very high indeed and far outweigh any potential revenue from other forms of recreation. Whilst this may explain why some estates are unwilling to allow the public access, even on a paying basis, to any part of the woodland (or farmland) the extent to which public access could in fact be compatible with shooting is often not decided on a rational basis.

References

1. H M Treasury. Forestry in Great Britain. An Interdepartmental Cost/Benefit Study. HMSO 1972.
2. *Forestry Commission. The First 75 Years. 1994.*
3. 50th CAR
4. 51st CAR
5. Forestry Commission. British Forestry 1974
6. 48th CAR
7. *Countryside Act (Scotland).* 1967.
8. 49th CAR
9. *Countryside Act.* 1968.
10. 54th CAR
11. *The Forest Resources of the Temperate Zones. The UN-ECE/FAO 1990 Forest Resource Assessment. Volume II. Benefits and Functions of the Forest.* United Nations 1993.
12. *Forestry Commission. Annual Report and Accounts 1993–1994*, Table 12.
13. *Forestry Commission Facts and Figures 1999–2000.*
14. *Forest Enterprise Annual Report 1999–2000.* Stationery Office Edinburgh 2001.
15. Grayson A.J. *Private Forestry Policy in Western Europe.* Scottish Forestry Trust 1993.
16. Hart, C. *Practical Forestry For the Agent and Surveyor.* Third edition, 1991.

Figure 6.1. Chain saw operator with protective clothing, 1980s

Chapter 7

The 1980s

The early 1980s was a time of general recession, deep enough to adversely affect many sectors of British industry, including the forest products industries using British timber as their raw material.

Against this unpropitious background, in 1980 the Government launched a major review of forestry policy from which the Forestry Commission emerged with credit [**1**]. It was able to demonstrate that it had led the greatest expansion of British forestry on record, forging a unique partnership with the private woodlands sector and aiding the private forest products industries to establish a whole range of new factories in many different parts of the country. Efficiency and productivity in the establishment of plantations and the harvesting of timber had been raised to levels previously undreamt of. Its public recreation policy had been an unqualified success and it was achieving new standards of amenity and nature conservation.

The case for forestry was strengthened by the timely publication in February 1980 of "A Strategy for the UK Forest Industry" by the Centre for Agricultural Strategy at Reading University [**2**]. This study called for a wider recognition of the importance of forestry to the national economy and emphasised the need for a more positive attitude to forestry. It concluded that there was a good case for further investment in UK forestry by both the public and private sectors in view of the forecast of increasing demand for wood and wood products and the prospect of future world shortages resulting in higher timber prices—conclusions which were broadly in line with the Commission's own, earlier, review "The Wood Production Outlook in Britain" [**3**].

1 Forestry Policy Statement

In December 1980 the Government issued a policy statement following on their extensive review of state and private forestry in Britain [**4**]. The Government pledged full support for a continuing expansion of forestry and its associated industries "in the National interest" and reaffirmed its intention to "continue the current income tax arrangements for (private) forestry in order to maintain confidence in the private sector" and to help maintain the successful and harmonious partnership between the private sector and the Forestry Commission which it regarded as "the main basis of policy for the future".

2 Dedication Basis III and Small Woods Scheme Closed to New Applicants; Forestry Grant Scheme and Broadleaved Grant Introduced

In July 1981, the long-term Dedication Basis III and the Small Woods Scheme were closed to new applicants and were replaced in October by an entirely new short-term (5 year) Forestry Grant Scheme (FGS) offering substantially higher levels of grant aid [**5**]. This scheme provided the stimulus to increased afforestation by the private sector. A further stimulus to private planting was given in 1985 when in October a special broadleaved grant was given to encourage the better management of broadleaved woodlands including the planting of broadleaves when re-stocking existing broadleaved woodlands, in new planting on sites suitable for broadleaves and, where appropriate, in re-stocking coniferous woodlands.

3 Forestry Act 1981—Sale of Forestry Commission Land

Increased Government "support for private enterprise" and therefore greater private participation in forestry also led to the Forestry Commission being directed to sell off substantial areas of its woodland estate. But the Government found that it had inadequate powers to dispose of Forestry Commission land and plantations and so a Forestry Bill giving the Government wide powers of sale was laid before Parliament.

There was widespread opposition to the sale of land policy by a variety of organisations, by very large numbers of the general public and by many in the House of Lords—including landowners with substantial forest interests.

One major objection by people from all walks of life was the likelihood that the purchasers of Commission woodlands and open land would stop all public access to them. In fact this fear has been fully justified and little can be done in practice to make such sales conditional on public access continuing.

Some 80 amendments to the Bill were tabled and it took up some 40 hours of debate, but eventually it passed into law as The Forestry Act 1981 [**1**].

The Bill also provided for the appointment of an additional Forestry Commissioner with business and commercial experience outside the forestry and wood using industries.

4 Forest Recreation

In their 60th Annual Report 1979–1980 [**6**] the Commissioners reported that they had had to make substantial reductions in their recreation budget over the past two years and that priority was being given to the revenue-earning developments such as campsites and holiday cabins and visitor centres and popular day-visitor facilities, particularly those close to towns.

5 The 1982 Valuation of the Commission's Estate—Second Quinquennial Revaluation

In their 62nd Annual Report [**5**] the Commissioners reported on the Second Quinquennial Revaluation of their estate as follows:

In reporting on the outcome of this latest revaluation of our estate, we feel that it would be helpful to explain the system of accounting under which the Forestry Commission has operated over the past ten years. Until 1972, it had been the practice to charge interest in the Commission's accounts at the current Government lending rate, irrespective of the earning power of the industry. Over the greater part of the Commission's existence, however, investments had been made for what were regarded as strategic and social reasons, without an expectation that a return at the current lending rate would be achieved. Against this background, and with most of the expenditure on the existing semi-mature forest estate having by then been incurred and the concept of a strategic reserve of timber for wartime use having ceased to have the same relevance, it was decided by the Government in 1972 that it would be opportune to restructure the Commission's accounts and to require in future that the Forestry Enterprise should aim at a target rate of return of 3% in real terms. Such a return was regarded as the maximum that might be earned from investment in afforestation in Northern Europe on the best sites close to markets. A new valuation of the Commission's plantations was determined by reference to an estimate of future net revenue discounted at the 3% rate, resulting in a substantial reduction in the figure taken as the value of capital employed at that time.

It was also decided that the accounts should show:

1. The difference between 3% and the rate of return generally required on Government trading accounts (currently 5%).
2. The extent to which new planting and restocking could not earn the target rate of return.
3. The cost of providing for recreation, amenity and nature conservation for which no commercial objectives could be set.

The purpose of identifying these elements as subsidies is three-fold. First, in order to show the subsidy to forestry arising from the Government's decision to set the target rate of return at 3% rather than the 5% generally required on commercial activities in the public sector; second, so that a periodic assessment can be made of progress in the achievement of the 3% target rate of return, as measured by the current valuation of the Commission's assets; and third, to show the costs attributable to certain non-commercial social objectives laid upon the Commission by successive Governments.

Planting and restocking subsidies are calculated as the capital sum required to bring the expected internal rate of return on a new plantation up to the 3% target. The forestry subsidy, on the other hand, is calculated as the difference between the annual return of 3% on the value of the forest estate as compared with a return of 5% on that value. The forest subsidy on a new or restocked plantation is therefore recorded as a note to the accounts and is spread over the whole length of the rotation rather than as a lump sum in the year in which the planting takes place.

Provision was also made for the calculation of standards for plantations expenditure and income and the assessed value of timber cut. The annual accounts from 1972–1973 onwards have shown the actual income and expenditure against these standards.

Under the new arrangements the Commission's estate has been revalued at five-yearly intervals since 1972. Each valuation has been based on a physical inventory of the growing stock and the fixed assets. A comparison of successive valuations is made to show how the investment in forestry has developed and whether progress is in line with the achievement of the target rate of return. It is also possible to see annually whether the programmes are being achieved and to judge approximately how far the departures from the price and cost standards necessary to earn 3% in real terms are attributable to general movements in the level of prices, and how price variations are balanced by those in costs.

The first quinquennium ended on 31st March 1977 and the results were set out in our 1976–1977 Report. The second quinquennium ended on 31st March 1982, and a further revaluation as at 1st April 1982 has been completed. The main conclusions to be drawn from it are as follows:

1. During the quinquennium 1977–1982, there was a shortfall against the standards set in 1977 for Plantations and the Harvesting and Marketing Accounts amounting to £61.6 million expressed in £ (1982). The main reason for this was that while cost was generally held within the standards, timber prices failed to keep pace with inflation as a result of the depressed market and exchange rate fluctuations. When account is taken of current assumptions about future work and output for the remainder of the rotation, using the 1977 cost and price assumptions updated by general inflation, the result is a further reduction of £89.1 million in the plantations value. The total shortfall against the cost, price and programme assumptions built into the 1977 valuation is therefore £150.7 million. Expressed as a return over the whole rotation, and after taking account of subsidies, this represents an overall return of 2.6% against the 3% target rate.
2. A reassessment of the plantation cash flows, based on updated estimates of future costs, prices and staffing levels to provide both the standards and the starting valuation for the next quinquennium, more than offsets the reduction of £89.1 million in the value of plantations previously mentioned. The total value of the Enterprise assets to be brought into the 1982–1983 accounts at 1st April 1982 is £1087.1 million, which is £0.2 million more than the figure in the Balance Sheet at 31st March 1982.
3. The target rate of return will remain at 3% for the 1982–1987 quinquennium as the internal rate of return on good sites close to markets is still expected to be at about that level. The actual rate of return in more remote areas can be expected to be below 3%, however, and we shall be looking for an average internal rate of return in respect of new planting and restocking during the period 1982–1987 of not less than 2.25%, compared with 1.75% in 1977–1982.
4. The forest recreation and amenity subsidy will be contained within a ceiling of £4.5 million in real terms for the 1982–1987 quinquennium.

6 The 1982 Forestry Commission Management Reorganisation

In 1982 the two levels of management—the District Officer in charge of a District and the Forester in charge of a Forest—were amalgamated [**7**].

The creation of a single tier of management below Conservancy level resulted in a saving of 20% of clerical and supervisory staff at that level. Subsequently, the number of Conservancies was reduced from eleven to seven—three in England, three in Scotland and one in Wales. The Deputy surveyorships of the Forest of Dean and the New Forest remained, but as the equivalent of Forest Districts.

Thus the Commission now operated with only three levels of management—Headquarters, Conservancy and Forest District—compared to five—Headquarters, Country, Conservancy, District and Forest—in the immediate post-war years.

In 1984 the Commission embarked on a major programme to exploit advances in computer technology, linking the three levels of management.

7 1985 Wildlife and Countryside Act; Balance between Timber Production and Conservation

In 1985 the Wildlife and Countryside (Amendment) Act [**8**] gave the Commission powers to balance timber production with conservation of features of special interest.

The following duties were included in this Act:

> "In discharging their functions under the Forestry Acts 1967 to 79 the Commissioners shall, so far as may be consistent with the proper discharge of those functions, endeavour to achieve a reasonable balance between:
>
> (a) the development of afforestation, the management of forests and the production and supply of timber and
>
> (b) the conservation and enhancement of natural beauty and the conservation of flora, fauna and geological or physio-graphical features of special interest."

In 1987 an EU directive required that an Environmental Impact Assessment be carried out where new planting (afforestation) might lead to adverse ecological change.

8 Forestry Industry Committee of Great Britain

In January 1987 the Forestry Industry Committee of Great Britain[1] was formed by private sector interests to represent all sectors of the forestry, timber and forest products industries. The Council's declared purpose [**9**] is to ensure that the en-

[1] In 1995 the title changed to Forestry Industry Council of Great Britain.

tirety of the national forestry resource is fully understood and properly presented to the Government, the media and the public.

In December the Council published a report entitled Beyond 2000. The significance of this publication is that it sets out for the first time the views of the private sector of the forestry and forest products industries as a whole.

In 1998 the Council with the assistance of the Forestry Commission published a Reference "yearbook" for the British Forestry Industry [**10**]—the most comprehensive source of data available. The intention is to update the information every two years.

9 Our Common Future 1987; World Commission on Environment and Development

In 1987, the World Commission on Environment and Development (The Brundtland Commission) [**11**] focused international attention on the health of the worlds forests and defined substainable development as development that meets the needs of the present without compromising the ability of future generations to meet their own needs. This is discussed in Chapter 8. See also Chapter 4 of the European Timber Trends Study published in 1986 [**12**].

10 The 1988 Budget Removal of Forestry's Tax Benefits

A major change in forestry policy was the removal in the 1988 Budget of the tax advantages which had been the main encouragement for private sector investment in afforestation and the long-term management of estate woodlands [**13**]. The shock to private woodland owners was all the greater because in the 1981 statement on forestry policy it had specifically endorsed the previously held views of successive governments—"The Government intend to continue the current Income Tax arrangements for forestry in order to maintain confidence in the private sector".

11 Woodland Grant Scheme Introduced

In March 1988 the Forestry Grant Scheme and the Broadleaved Woodland Grant Scheme were closed to new applicants and in April 1988 were replaced by a single scheme—The Woodland Grant Scheme (WGS) [**13**] which ran from June 1988 to June 1991, and was entitled WGS Mark I.

The day following the Budget, the Secretary of State for Scotland as the lead Forestry Minister sought to re-assure the forestry industry that the changes in support for private forestry did not mean that there had been any fundamental change in the Government's policy of encouraging forestry. The Forestry Commissioners reported that these assurances were welcomed by the forestry industry. It would have been strange if they were not!

Table 7.1 New planting (Afforestation, in 1000s ha) by the Forestry Commission and by private woodland owners in years ending 31 March

Year	FC	PW	Year	FC	PW
1981	11.6	8.7	1986	4.3	19.4
1982	11.0	12.6	1987	5.3	19.4
1983	9.0	12.5	1988	5.0	24.0
1984	8.4	16.9	1989	4.1	25.4
1985	5.2	16.6	1990	4.1	15.6

Referring to the increase in private planting in the decade up to 1989 the official Forestry Commission historian comments [**1**]:

> "This tremendous achievement was the more remarkable for the fact that it took place against a background of rapid change. Over a period of ten years, four grant schemes were closed and three were introduced."

12 Ban on Afforestation in English Uplands

The Secretary of State for the Environment and the Minister of Agriculture, Fisheries and Food in 1988 announced a ban on the creation of new woodlands in the English Uplands where site conditions were favourable only to conifers [**13**].

13 Farm Woodland Scheme

The Farm Woodland Scheme opened to applications in October 1988 [**14**].

Its objectives included:

- Reducing agricultural production (much of it surplus to the country's needs).
- To help compensate for the continuing loss of rural employment as agriculture became more and more mechanised.
- To contribute to farm income, a long-term aim in new forestry schemes where financial rewards follow some 20 to 80 years after planting, but here to be subsidised by annual grants to farmers who planted farmland.

14 Progress of Sale of Forestry Commission Woods and Land

By the end of the decade, more than £120 million worth of Commission forests had passed into private ownership and—as predicted by critics of the policy in 1981—many of these woodlands were closed to public access by their new owners, giving rise to increased opposition to the disposals policy by large numbers of people who had hitherto enjoyed their right to fresh air and exercise in Commission woodlands.

15 Community Forests

A programme to create community forests on the outskirts of major cities and towns in England and Wales was launched jointly by the Forestry Commission and the Countryside Commission in July 1989 [**15**].

The 70th CAR reports:

"A national programme to create community forests on the outskirts of major cities and towns in England and Wales was launched jointly by the Countryside Commission and ourselves in July 1989. Each community forest could cover between 10 000 and 20 000 hectares managed for a wide variety of recreational, environmental and commercial uses. The long-term aim is to create a mixed woodland setting with open spaces and farmland. Community forests will help restore derelict and unkempt parts of the urban fringe and green belt, as well as providing new jobs in forestry, leisure and conservation.

We are arranging for a Forestry Commission forester to serve as a member of each of the project teams for the three community forests already approved by the Government in South Staffordshire, South Tyne and Wear and East London. In addition to providing grants for tree planting through the Woodland Grant Scheme, we shall be giving assistance in the form of training and research and specialist advice, for example on landscape design and forest recreation.

The emphasis in planning the community forests is to involve local communities, local authorities, local businesses, voluntary bodies and others in the design, establishment and long-term maintenance of the forests. As lead partners at a national level, the Forestry Commission and the Countryside Commission will steer the programme and provide grant aid.

Local authorities will be key participants—contributing funds, expert advice and staffing—and it is hoped that the projects will also attract substantial sponsorship and practical support from local industry and commerce.

Each forest will be co-ordinated by a project team, guided by a steering committee of representatives from the major partners and local authorities, and will prepare a community forest plan in consultation with the other interests. Project team members for the first three forests were expected to be appointed in the spring of 1990.

This is a bold and ambitious programme, which will require the sustained effort of many different organisations and local communities. We shall do all we can to support and encourage it."

16 British Forest Products Industries

But what about the forest products industries in the 1980s?

Early in the decade, three of the paper pulp mills using British timber closed. These were the Wiggins Teape chemical pulp mill near Fort William [**4**], built in the early 1960s to utilise Scottish conifer thinnings to make bleached pulp for fine papers; Bowaters groundwood mill at Ellesmere Port [**4**], where since 1958 mechanical pulp for newsprint had been made; and the St. Anne's thermo-mechanical pulp mill in Bristol, making pulp for cardboard carton manufacture [**4**].

The forest products industry and especially the pulp and paper sector competes on a world-wide basis and production in every country is vulnerable to exchange rate fluctuations and other external factors outwith its control.

The problems that beset the relatively young British industry—serious though they were for individual firms—were seen by many as short-term in nature. To encourage investment in Britain—one of the few countries with a rapidly increasing production of coniferous roundwood from a sustainable forest resource—the Forestry Commission took the lead in a search for new investment to make up for the loss in demand for pulpwood caused by the mill closures referred to above and in the immediate term to find export markets. A Scottish Forest Products Development group was formed to provide potential investors with a one-stop source of information on wood supplies and financial assistance [**1**]. Within 5 years some £600 million of new investment had been attracted and the British paper pulp and panel products industries had taken on an entirely new lease of life.

The sawmilling industry was at one time in the hands of long-established family firms serving local or special markets. Even during the Second World War the vast increase in production came about by increasing the numbers of small local sawmills rather than by building larger sawmills. The main reason for this was the low density of woodlands in Britain, which meant that large sawmills would have had to draw their logs from a wide radius and so incur high transport costs. Another important factor was that unlike the other sectors of the forest products industry, in the immediate post-War years, there were no major technical innovations in sawmilling leading to the great "economies of scale" so evident in the pulp and panel products sectors.

As British sawmillers began to see a rise in the potential supplies of sawlogs from the large areas of coniferous plantations created in the immediate post-War years, they followed closely the new developments in sawmilling methods that were slowly evolving in Scandinavia and North America from the late 1960s through the 1970s. These new methods were made possible by advances in basic and applied science which could be incorporated in manufacturing processes and included the use of computers linked to electronic scanning devices that replaced the sawmiller's experienced judgement as to how to set the machinery to maximise the sawn outturn from the wide assortment of qualities of log brought mechanically to the saws.

Table 7.2 Analysis of deliveries of timber produced in Great Britain to sawmills and wood processing industries

	Thousands m^3 of under bark									
	1980	1981	1982	1983	1984	1985	1986	1987	1988	1989
Softwood sawlogs	1640	1670	1865	1875	1825	2440	2390	2570	2855	3045
Hardwood sawlogs	940	905	715	580	535	520	650	440	475	665
Round pitwood	145	145	130	120	45	100	90	65	45	35
Chipboard[a]	195	215	180	245	285	370	635	735	990	955
Fibreboard	25	30	35	30	15	20	20	15	20	15
Paper and paperboard	600	855	830	725	770	945	1085	1215	1180	1215
(of which exports)	(210)	(590)	(555)	(430)	(350)	(355)	(320)	(380)	(255)	(160)
Other industrial wood	285	275	130	130	195	155	145	155	160	175
TOTAL INDUSTRIAL WOOD	3830	4095	3885	3705	3670	4550	5015	5195	5725	6005

Source: Forestry Facts and Figures published annually by the Forestry Commission.
[a]Chipboard = particle board.

Advances in the manufacture of saws and in log handling equipment generally speeded up the whole movement of logs and sawn wood through the sawmill to the sawn wood grading tables and mechanised stacking devices.

The above Forestry Commission figures (Table 7.2) show how well the forest products industries were able to overcome their problems during the 1980s.

The 1989 timber production figures show the remarkable increase in the production of softwood (coniferous) sawlogs from 1.6 million m^3 in 1980 to 3 million m^3 in 1989, an increase of 86%. Softwood sawlogs rose from 42% of total production to 50% in the same period. This was due not only to the increased availability of sawlog size material from the woodlands but also to the enormous strides made by sawmillers in modernising their sawmills to produce accurately sawn, seasoned (dried) and graded lumber to meet the precise specifications of the main user of sawn softwoods, the building industry. At last the sawmilling sector was competing on equal terms with most modern mills in Scandinavia, Canada, Russia and other traditional suppliers of Britain's building timbers.

By 1989 demand for round pitwood for use in coal mines, once the main market for small-sized softwood from thinnings, had dwindled to less than 1% of total demand for British timber. Its place was taken by the chipboard (particle board) and fibreboard industries—with their less exacting specifications—and the paper and paper board industries which are able to pay substantially better prices than the board industries even allowing for the fact that they require a higher quality of wood.

17 Indicative Forestry Strategy Plans

The Secretary of State for Scotland announced in February 1989 [**14**] that he would be encouraging Regional Councils to prepare indicative forestry strategies as part of their structure plans, to include the publication of a strategy map to show 'preferred', 'potential' and 'sensitive' areas for new planting.

Early steps in this general direction were taken in the mid to late 1970s, but this was a much more detailed and positive approach than the early pioneering work. As already noted in Chapter 5, the Macaulay Institute's land capability for forestry maps covering all of Scotland were an important basis for planning.

18 The European Union and Forestry

A major step was taken in the recognition of forestry by the European Union in 1989 when it set up in Brussels a Standing Forestry Committee [**15**].

The Standing Committee will, for the first time, provide a means for the forest services of Member States to influence proposals affecting forestry at a very early stage.

The EU view is that there is a need for the expansion of forestry within the Community which is only some 50% sufficient in wood and the UK will benefit from financial aid towards new planting on farm land under the Forestry Commission's Woodland Grant Scheme.

References

1. *The Forestry Commission. The First 75 Years.* 1994
2. Centre for Agricultural Strategy. Reading University. *A Strategy for the UK Forest Industry.* 1980
3. *The Wood Production Outlook in Britain.* Forestry Commission. A Review 1978
4. 61st CAR
5. 62nd CAR
6. 60th CAR
7. 63rd CAR
8. 66th CAR
9. 67th CAR

10. *The Forestry Industry Council of Great Britain, A Reference For the Forestry Industry.* FIC GB (a), 5 Dublin Street, Edinburgh EH1 3PG
    Note (a): GB later dropped from the title
11. Forestry Commission. *The UK Forestry Standard.* Edinburgh 1998 (page 8).
12. United Nations. *European Timber Trends and Prospects to the Year 2000 and beyond.* Chapter 4. E.G. Richards et al. New York 1986
13. 68th CAR
14. 69th CAR
15. 70th CAR

Chapter 8

The 1990s

1 Introduction: International Interest in Forestry and the Environment

This brief introduction to the 1990s refers specifically to the international initiatives following the 1987 Brundtland Commission report "Our Common Future" which focused attention on the health of the world's forests, on climate change and other globally significant environmental and development issues and introduced the concept of sustainable development. These initiatives are discussed more fully later in the chapter, but it is thought useful to list them here in summary form.

In 1992 at the United Nations Conference on Environment and Development (UNCED) in Rio de Janeiro (The Earth Summit) world leaders adopted a number of agreements relevant to forests, notably a Statement of Forest Principles, Agenda 21; and at a ministerial conference in Helsinki in June 1993 European Governments adopted a set of guidelines for European forests, including sustainable management and conservation of their bio-diversity.

Following the 1993 agreement at Helsinki, which included General Guidelines for the Sustainable Management of Forests in Europe and General Guidelines for the Conservation of the Bio-diversity of European Forests, the UK published in 1994 Sustainable Forestry—the UK Programme (Cm 2429) [**1**].

Then in 1994 European countries proposed a set of Pan-European Criteria (PEC) to be applied to forests throughout Europe, whatever their location or circumstance.

In 1995, the UK Government decided to publish a UK Forestry Standard which would bring together in one document the criteria and standards for sustainable management of forests in the UK. After two public consultations, in 1996 and in 1997 and bilateral consultations with other Government Departments and major Non-governmental Organisations (NGOs) the document was published in 1998, as The UK Forestry Standard. The Government's Approach to Sustainable Forestry. Edinburgh: Forestry Commission [**2**].

In May 1998 the G8 Birmingham (England) Summit [**3**] and in March 1999 FAO in Rome, produced guidelines and declarations on forestry policy in relation to sustainability, environment and development [**4**].

Whilst these broad issues were being debated at international level, the British Government undertook a series of reviews of forestry policy in Great Britain, including regulatory and grant-aid schemes for private forestry, where appropriate taking into account policies and guidelines incorporated in international agreements.

At the time of writing, the full impact of the various initiatives of Government and the Forestry Commission on British Forestry is unclear and uncertain.

2 Forest Policy Reviews 1990–1994

2.1 Broadleaves Policy Review 1990

On 23rd July 1990 the then Secretary of State for Scotland in his capacity as Forestry Minister, made an announcement in a Written Answer to a question about the outcome of the Government's review of its policy for broadleaved woodland, following extensive consultation with a wide range of forestry and nature conservation interests.

The following extracts are taken from Appendix VI of the Forestry Commission 71st Annual Report and Accounts 1990–1991 [**5**].

"The Government's policy for broadleaved woodland was announced on 24th July 1985. The main objective of the policy was described in the statement as being 'to maintain and enhance the value of Britain's broadleaved woodland for timber production, landscape, recreation and nature conservation'.

Several new measures designed to achieve the policy aims were introduced or foreshadowed by the 1985 statement (and these included) the introduction of a Broadleaved Woodland Grant Scheme designed to assist in the rehabilitation of neglected woodland and to encourage the creation of new multi-purpose broadleaved woods. The main features of the Broadleaved Woodland Grant Scheme were incorporated in the Woodland Grant Scheme following our decision to remove forestry from the scope of income tax and corporation tax in 1988.

The Government accepts that more needs to be done to ensure the better management of our broadleaved woodlands in the future, to encourage owners to bring neglected woodlands into a healthy and productive state, and to provide owners with the necessary incentives to maintain and improve the environmental value of their woodlands. We have therefore decided to make woodland management grants available under the Forestry Commission's Woodland Grant Scheme. In the light of the changes to the forestry tax arrangements and the considerable scope which exists to improve the environmental value of all types of woodland, we have further decided that woodland management grants will be made available for conifer as well as broadleaved and mixed woodlands.

Subject to clearance by the European Commission under the terms of Article 93 of the Treaty of Rome, this significant extension to the Woodland Grant Scheme will come into operation on 1st April 1992, with the first grants being paid in 1993–1994.

Under the present provisions of the Woodland Grant Scheme, planting grants are paid for the initial establishment of new coppice stools but only on the basis of traditional coppice rotations. We have decided that this is unnecessarily restrictive and the Forestry Commission will be introducing an amendment to the Scheme so that the establishment of short rotation coppice will be eligible for planting grants, subject of course to the silvicultural and environmental conditions of the Scheme being met. Coppice woodland can have particularly high conservation value if managed in a traditional way and will be eligible for the woodland management grant provided that the other conditions are met.

The Government is grateful to all those who contributed so constructively to this review of broadleaves policy, and we are sure that the decision to introduce woodland management grants will be particularly welcomed by both forestry and environmental interests.

The policy we introduced in 1985 has proved to be soundly based and has already had a significant impact on British forestry. With the changes I have announced, we are well placed to continue with the important task of conserving and enhancing the value of our woodland heritage."

2.2 The 1991 Forestry Policy for Great Britain

On 18th September 1991 the three forestry ministers—published a "Forestry Policy for Great Britain" [**6**] in order to draw "the various strands of forestry policy together into a single statement". They claimed that the "record of Government forestry policy initiatives over the past 12 years (i.e. since 1979 when the Forestry Commission celebrated its Diamond Jubilee) has been one of considerable progress and achievement".

One can be forgiven for thinking how much the Forestry Commission, private owners and the forest industries achieved in this period in spite of the multiplicity of changes in grant aid schemes imposed by Government.

Although the short introduction to the publication mentions wood supply first, the main text on Aims lists the benefits of the forests and woodlands in the following order:

- Recreation and public access to the countryside.
- A valued component of landscape.
- An essential habitat for wildlife.
- An increasing contribution to the growing national demand for timber.
- Rural employment.
- An effective means by which carbon dioxide can be absorbed and stored over long periods of time.

The two main aims of the Governments forestry policy are thus:

- The sustainable management of our existing woods and forests.
- A steady expansion of tree cover to increase the many, diverse benefits that forests provide.

It is a pity that sustainable timber production is not mentioned; without that as a top priority the nation could not afford to create and manage woodlands to meet the other objectives.

The priorities were better stated in the mid-1980s when in the Wildlife and Countryside Amendment Act (see Chapter 7) the Forestry Commission was directed by the Government to balance its sustainable production of wood policy with the environmental benefits—amenity, recreation and nature conservation—that the forest can provide. Some would say that this directive was no more than a recognition of the way that the Forestry Commission had of its own accord moved on from its earliest mandate "the nation needs timber" towards a broader-based more environmentally-sensitive approach. Others would argue that the directive gave a much needed prod to the Commission to be much bolder in pressing the case for multiple objective forestry in face of Treasury preference for the efficient production of timber as the over-riding objective. Whichever view one takes, the fact is that the early 1990s heralded in a time of more generous Government grants to private owners for environmentally friendly forestry and a Forestry Commission being allowed to spend substantial sums on landscaping and nature conservation and to achieve other environmental benefits in a way and on a scale that would earlier not have attracted Treasury approval.

On 25th July 1991 a Bill, the Forestry Act 1991 [**7**], received Royal Assent. Its purpose was to increase the maximum membership of the Regional Advisory Committees from nine to twelve so as to allow a wider representation of environmental interests on the Committees. Previously the membership was weighted in favour of forestry since the Forestry Act required that four out of the nine members were to be appointed after consultation with forestry interests.

The Forestry Commissioners, in their 71st Annual Report, said that they would fill the extra three places with persons from recognised environmental backgrounds; that they greatly welcomed the widening of the Commission's expertise; and that they hoped that as a result this would lead to a greater understanding and acceptance of their work [**8**].

2.3 The 1991 Crofter Forestry (Scotland) Act

The 1991 Act came into force on 1st April 1992 [**9**]; it permits crofters grazing committees to establish woodland on their common grazing land. This is an important change since *inter alia* substantial areas of potential planting land in the crofting counties of Scotland has been the common grazings, not normally available to forestry for historic legal reasons.

> The legal recognition of Scottish crofting as a form of land tenure goes back to the 19th century; in 1886 the Government acted to end the injustices of the "clearances" of the 18th and 19th centuries when many Highland landlords ruthlessly cleared their agricultural tenants from relatively fertile land they had long cultivated—to exile overseas or to poor infertile land often near the coast—in order to make way for sheep and to a lesser extent for deer stalking.

None of the Forestry Commission's grant aid schemes for private planting really addressed the needs of crofters and as mentioned above, there were legal constraints which discouraged and often prevented crofters from engaging in forestry.

The Report of a Commission of Enquiry into Crofting Conditions in 1955 [**10**] took the line (36th CAR) that State forestry schemes in the Scottish Highlands would provide crofters with employment additional to crofting activities and the Forestry Commission helped by the Scottish Department of Agriculture undertook to try to increase their acquisitions of plantable land in the Scottish crofting counties to increase employment opportunities for crofters, "even if this means in some instances planting on land which will give a smaller return than it (the Commission) normally looks for". The 1991 Crofter Forestry (Scotland) Act was the first ever piece of legislation to tackle the question in a realistic way making it possible to use the various forestry grant schemes to create new croft-based woodland, with broad environmental benefits as well as or in place of the once mandatory timber production requirement of all previous Forestry Commission grant aid schemes.

Because of long-term implications of the Act to the long-term future of land use in north and north west Scotland, the question is discussed more fully in Chapter 9.

2.4 Management Re-Organisation 1992

In 1966 the Commission's accounts had been re-structured [**11**] and the terms Forestry Authority and Forest Enterprise were used to distinguish between:

- the Commission's functions as a Government Department and
- the Commission's function as a trading body managing the country's publicly-owned forests.

In April 1992 the Forest Enterprise (as it was now named) became a self-contained entity within the Forestry Commission and the Forestry Authority became a self contained arm within what was to be called the Department of Forestry [**12, 13**].

The Forestry Authority, now totally freed from the day-to-day management of the State forests, was made responsible for the implementation of forestry policy (including approval of State forestry plans); the control and supervision of private forestry, including the grant aid schemes; the various regulations which apply to tree felling and to the control of pests and diseases; and public information for informing the public about all aspects of forestry and its potential as a natural renewable resource.

The duties of the Commission as the Department of Forestry in Britain were strengthened by the creation of a Policy and Resource Group. Among its tasks are the provision of support to the Forestry Ministers on Parliamentary matters and in the development and monitoring of forestry policy. It is also responsible for international forestry questions and the provision of central support services for the Commission as a whole in such areas as computer support, personnel, internal auditing and public information.

The Forest Enterprise is the organisation which manages the Commission's forests, including forest recreation and its non-forest estate.

2.5 The 1993 Forestry Review Group

On 30th March 1993 the Secretary of State for Scotland, announced a review of both the effectiveness of forestry investment incentives and the ownership and management of Forestry Commission woodlands [**12, 13**].

A year later, The Forestry Commission's house journal reported thus:

> While the result has been a year of uncertainty for Commission staff, we've been able to take heart from the almost universal support our work

has received—sometimes from organisations that have been our bitterest critics in the past.

Privatisation rumours brought an immediate reaction from conservationists, with organisations such as the Council for the Protection of Rural England and the Campaign for the Protection of Rural Wales speaking in favour of the Commission almost immediately. Local groups also expressed concern and in April 1993 the RSPB, backed by 850 000 members, joined the Commission's defence. Professor David Bellamy, a leading environmental activist, threatened to chain himself to the gates of Buckingham Palace if privatisation went ahead. "Our national forests are a key resource for conservation and amenity—a green relief valve without equal", he said in November.

Public access was an issue from the beginning. The Ramblers' Association announced a campaign in April and more than 10 000 walkers turned out at 100 sites in September to protest against the threat to access.

Even the private forestry sector largely backed the status quo. In July, Timber Growers UK told the Government "there is no need to privatise the Forestry Commission". The Duke of Buccleuch, one of the country's largest private landowners, said "It would be crazy to dismantle or dismember (the Forestry Commission) at such a time. Unless we succeed in this task our children and grandchildren will have every reason to reproach us".

The wood processing industries were equally opposed.

And strong opposition to the Conservative Party's views came from Labour, Liberal Democrats and Welsh and Scottish Nationalists. In part, this opposition was fuelled by speculation on job losses, particularly in Scotland and Wales.

When the Review Group reported to Ministers in March 1994, it had been bombarded with more than 3000 letters and received representations from 300 interest groups concerned about everything from rights of access to conservation.

In July 1994 the Secretary of State for Scotland announced in the House of Commons [**14**] that:

"Forestry Commission woodlands will remain in the public sector; Forest Enterprise will be put on a more business-like footing as a Next Steps Agency headed by a Chief Executive;

Access arrangements (to Private Woodlands) are to be expanded and strengthened;

New planting incentives will encourage private sector planting of productive conifer forests".

3 The Progress of State (Forestry Commission) and Private Forestry 1920–1994

3.1 Forestry Commission

Table 8.1 Forestry Commission planting (in thousands of ha)

Years	Afforestation		
	New planting	Re-stocking	Total
1920–1948	160	59	219
1949	12	2	15
1950–1959	147	98	245
1960–1969	163	49	213
1970–1079	161	33	194
1980–1989	80	67	147
1990–1993	13	32	45

See also Figure 8.1. Totals do not always add up due to rounding.

3.2 The Progress of Private Forestry 1920–1994

In spite of all the uncertainties in Governments forest policies, the Secretary of State could well be pleased with the progress of the development of the role of private woodlands since 1950.

While the progress up to that date was modest, private woodlands planting, afforestation and re-stocking over the years amounted to 69 338 ha: 1920–1929 29 483 ha, 1930–1939 21 450 ha and 1940–1948 18 405 ha. Subsequent progress is shown in Tables 8.2 and 8.3, which set out planting by the various grant-aid schemes over the years 1949–1971 (Table 8.2) and 1972–1994 (Table 8.3) which may be summarised (together with the latest overall figures 1995–1999) as follows:

Period	Total planting (ha)	Scheme
1950–1954	33 000	Dedication I+II from 1949
1955–1959	60 000	
1960–1964	71 000	
1965–1969	58 000	
1970–1974	114 000	Dedication I+II closed from 1972
1975–1979	61 000	Dedication III from 1976
1980–1984	73 000	Forestry Grant Scheme from 1983
1985–1989	126 000	
1990–1994	106 000	Woodland Grant Scheme from 1990
1995–1999	110 000	

Table 8.2 G.B. Planting in Private Woodlands—afforestation and re-stocking, with and without grant aid. Figures converted from acres to 1000s ha, total rounded to nearest 1000 ha

ye 30 Sept	Total	Grant aided		Small woods	Without grants	Other
		Dedicated approved	Other			
1949	5	—	4	—	1.2	
1950	7	0.4	5.7		1.2	
51	5	1.2	0.8	0.4	2.4	
52	6	3.2	0.4	0.8	1.6	
53	7	4.8		1.2	2.0	
54	8	5.3		1.2	1.2	
55	9	7.3		0.8	0.8	
56	11	8.1		1.6	1.2	
57	13	10.1		1.6	1.2	
58	13	10.1		1.6	1.2	
59	14	11.3				2.8[a]
1960	15	12.1				2.8[a]
61	15	12.5				2.4[a]
62	14	11.7		1.6	0.8	
63	13	10.9		1.2	0.8	
64	14	11.3		1.6	0.8	
65	13	10.5		2.0	0.8	
66	13	10.9		1.6	0.4	
67	15	12.9		1.6	0.4	
18 months to 31.3.69	17	14.9		1.6	0.4	
ye 31 March						
1970	19	17.4		1.6	0.4	
71	23	21.4		1.2	—	

[a]Planting in small woods and without grant aid not given separately for 1959, 1960 and 1961.
ye = year ending.

Table 8.3 GB private woodlands planting, afforestation and re-stocking with and without grant aid. Figures published in ha in CAR's. Total rounded to nearest 1000 ha.

ye 31/3	Total+	Grant aided			Small woods	Without grants	Other	
		Dedicated[a] approved	Basis III	Other				
1972	24	23			1	0.5		
73	24	22			1	1		
74	24	19		4[d]	0.5	1		
75	21	13		8[d]	0.1	0.4		
76	12	7	4	0.3[d]		0.4		
77	9	4	5			0.4		
78	8	3	5			0.3		
79	11	3	7		1[b]	0.3		
1980	11	3	7		1	0.5		
81	11	3	7		1	0.5		
82	16	15[c]			1	0.3		
83	15	7[c]		8FGS	0.2	0.5		
84	20	4[c]		16FGS		0.5		
85	20	3[c]		16FGS		1.0		
86	23	2[c]		21FGS		0.7		
87	24	2[c]		20FGS		0.4	BGS	2
88	29	2[c]		24FGS		0.3	BGS	3
89	30	2[c]		25FGS		0.2	BGS	3
1990	19	1[c]		5FGS 12WGS			BGS	1
1991	19	1[c]		2FGS 16WGS			BGS	0.4
1992	21	1[c]		1FGS 16WGS 3FWS[e]			BGS	0.5
1993	23	0.5[c]		0.3FGS 23WGS		0.2	BGS	0.1
1994	24	0.3[c]		0.3FGS 24WGS		0.2	BGS	0.1

[a]Dedicated Basis I and Basis II; also Approved Woodlands which closed June 1972 but existing participants allowed to complete their Plans of Operations. No AW planting after 1978.
[b]Small woods planting grant re-opened; incorporated in FGS 1981 onwards but some small woods continued to 1983.
[c]All Dedication Schemes.
[d]Interim Planting Grants Scheme
[e]Farm Woodland Scheme FWS in association with WGS.
Basis III: Dedication Basis III. Closed to new applicants July 1981.
BGS: Broadleaved Grant Scheme.
FGS: Forestry Grant Scheme started October 1981.
WGS: Woodland Grant Scheme MK 1.
1993 and 1994 WGS includes Farm Woodland Scheme and Farm Woodland Premium planting administered by Agricultural Departments.
The Interim Planting Grant Scheme was introduced in April 1973 as a bridge to Basis III from discontinued Basis I and II, Approved Woodland and Small Woods Planting Grants. It covered approved planting started and completed during the period 28th June 1972 to 30th September 1974 on land not included in any of the former grant schemes. The Scheme was closed to further applications from 1st January 1974.

Table 8.4 Forestry Commission grant schemes area under management (from Commissioners Annual Reports) in hectares

	1950	1960	1970	1980	1990	1994	(No. of Schemes in 1994)
Dedication Basis I&II	14 860	251 702	398 950	484 644	341 479	198 931	(1802)
Dedication Basis III				44 012	84 256	47 461	(442)
Forestry Grant Scheme					203 485		(NIL)
Broadleaved Woodland Grant Schemes					45 961		(NIL)
Woodland Grant Schemes					148 986	553 538	(20 692)
WGS+FWS[a]					7761		(NIL)
WGS+FWPS[b]						21 908	(3015)
	14 860	251 702	398 950	528 656	831 528	821 838	(25 951)

[a]Woodland Grant Schemes associated with Farm Woodland Scheme
[b]Woodland Grant Schemes associated with Farm Woodland Premium Scheme
No figures are shown for areas managed under the Forestry Grant Scheme, the Broadleaved Woodland Grant Scheme and the Farm Woodland Scheme in 1994. All plans of operations under these schemes have expired.
These figures show that in 1994 just over 60% of the total G.B. private productive woodlands area of some 1.3 million hectares was managed under a grant aid scheme with a plan of operations approved by the Forestry Commission.

The rate of new planting (afforestation) in Private woodlands was relatively stable in the early 1990s at around 15 000 ha a year—the Forestry Commission had by 1994 virtually ceased to expand by new planting (1400 ha in 1994).

Year	PW	000's ha
1991	15.4	
1992	14.1	
1993	15.3	
1994	15.9	

From Ref. [**14**].

4 Other Developments

4.1 The Forestry Research Co-ordination Committee (FRCC)

*The Forestry Commission Annual Report for 1993–1994 records the new style Forestry Research of the 1990s [**15**]*

Introduction

The Forestry Research Co-ordination Committee (FRCC), which brings together all the Government Departments and organisations sponsoring research on forestry, met four times during the year. The Research Users' Forum, established last year, met on two occasions. The Forum represents the timber industry, farming, land use, environmental and conservation interests and has assisted FRCC through identifying research priorities concerned with four topics: land use; low-intensity silviculture; deer and squirrel research needs; and marketing of low quality broadleaves. The Forum fulfils an important part of the technology foresight process. The Forestry and Conservation Working Group—a sub-committee of the FRCC—helps to co-ordinate research and identify priorities by bringing together senior scientists from the Institute of Terrestrial Ecology, the Forestry Commission and English Nature (also representing the interests of Scottish Natural Heritage and the Countryside Council for Wales).

Overall Expenditure on Forestry Research

Nearly £21 million was spent on forestry and related tree research in Britain in 1993–1994. This is a small increase over 1992–1993 following the inclusion of projects funded by agencies not previously included in the annual collation of research. Research on environmental and conservation aspects continues to attract increased support. For the first time since FRCC began collating research in 1982, half of all research in 1993–1994 concerned broadleaved tree species, predominantly oak, ash, beech, birch, sycamore, poplar and willow. Research linked to new woodland initiatives, notably planting of farmland and community forests, continues to play an important role.

Methods of Operation

FRCC continues to operate through the appointment of Review Groups consisting of acknowledged experts in the particular field of research. Five years after the publication of the initial Review Group Report, a follow-up Action Report is produced detailing the action or lack of it which has taken place. This procedure has served FRCC well but the Committee now finds that the process is becoming less productive in certain areas where the research is long-term. This consideration led to the formation of standing Working Groups on new areas of interest such as farm forestry, poplar cultivation and the general area of forestry and nature conservation.

New Systems

This new approach has enabled FRCC to be more pro-active. It has led, for example, to the creation of a special topic research fund for farm forestry and a conference to discuss the results of the sponsored research. A high-profile conference on carbon storage by trees is also being arranged for November 1994 to take forward the Review Group Report on that subject.

FRCC Users Forum

A very successful Research Users' Forum has been introduced which has set up Working Groups to produce recommendations on areas of research which users feel required to be looked at. These Working Groups have produced some very stimulating reports which will be discussed by the full Users' Panel before final presentation to FRCC.

New Constitution

The FRCC has during the past year discussed whether to reconstitute itself as an Inter-Agency Committee similar to those created for Global Environment Change (IGEC) or for Marine Science and Technology (IACMST). On balance, it was concluded that this change of constitution would not materially improve its ability to co-ordinate forestry research, so no further action was taken.

4.2 The June 1992 United Nations Conference on Environment and Development (UNCED), Rio de Janeiro

In the early 1990s Governments began to signal their increased awareness of the need for greater protection of the environment across a broad range of human activities. In June 1992 over 100 Heads of State or Government took part in a conference held in Rio de Janeiro under United Nations auspices. The conference confirmed the importance of environmental issues on the political agenda and produced a number of historic agreements [**12, 13**] including the Climate Change Convention (concerning global warming, e.g. emission of greenhouse gases and their absorbtion by forests which reduce the greenhouse effect) and the Convention on Biological Diversity which aims at the protection of plant and animal species and their habitats.

Of more direct importance for forestry was a Statement of Forest Principles, the first political agreement to apply to all the world's forests. The G7 and European Union countries have committed themselves to keeping the Statement of Forest Principles under review. As a result of an initiative by the UK, all G7 and EU countries made commitments to publish national plans for implementing the Statement of Forest Principles adopted at the conference.

4.3 United Kingdom Sustainable Forestry Programme

In January 1994 the Government published a booklet Sustainable Forestry the UK Programme [**1**] describing how the Government control of forestry in Great Britain and Northern Ireland ensures that UK forests are managed responsibly with proper regard for the environment and in accordance with the principles of sustainable forestry.

4.4 Timber Certification and the UK Woodland Assurance Scheme

International bodies (e.g. the Timber Committee of the Economic Commission for Europe) and Environmental organisations are becoming more and more interested in the certification of the origins of timber sold on the world markets. They wish to avoid so far as possible the felling and destruction of natural forests by timber traders for short-term gain and to try to ensure that only timber from sustainably managed forests—whether natural or planted—is traded on world markets. For some time the Forestry Commission has been trying to bring together the Forest Stewardship Council's[1] own draft standards and the Commission's UK Forestry Standard to develop a protocol for the certification of UK forests that has the support of the UK forestry and environmental bodies; takes account of the needs of growers; satisfies the demands of the timber industry's customers; and has international credibility.

Owners who manage their forests in accordance with the agreed published standard (see The UK Forestry Standard 1998, below) will be able to have their forests certified and this certification will mean that the timber from their forests can be sold on the domestic and world markets as coming from environmentally friendly forests managed on a sustainable basis.

It is appropriate to record here that in the Forestry Commission's Annual Report and Accounts for the year 1998–1999 [**16**] the Commissioners reported that (by March 1999) the woodland owners; woodland managers; wood processing, wood products, trade and environmental and social interests; were able to agree on the text of a Forest Certification scheme, The UK Woodland Assurance Scheme, that:

- has the support of the UK forestry and environmental communities
- takes careful account of the needs of growers
- satisfies the demand of customers
- has international credibility

In passing, it is interesting to note that in their 1986 European Timber Trends Study [**17**] the Timber Committee (TC) of the Economic Commission for Europe (ECE) and the European Forestry Commission (EFC) of the Food and Agriculture

[1]The Forest Stewardship Council, Unit D, Station Building, Llanidloes SY18 6EB, UK.

Organisation (FAO) included for the very first time a chapter on the Non-Wood Benefits of Forestry (nature conservation, amenity, recreation and products such as the many medicines that certain forests produce naturally). This was a recognition of the growing interest expressed by environmental organisations in the management of forests for these wider objectives and not only for sustainable wood production.

5 1995–1999 Forestry Commission Reorganisation

During the period 1995–1999, the Forestry Commission adopted a new form of words to describe its policies, aims and objectives; these became the Mission of the Forestry Commission and its Objectives, and were first published in the Forestry Commission Annual Report and Accounts for 1996–1997 [**18**].

5.1 Mission and Objectives

The Mission of the Forestry Commission is to:

Protect and expand Britain's forests and woodlands and increase their value to society and the environment

Our (Forestry Commission) Objectives as set out in Sustainable Forestry—the UK Programme 1994 are to:

Protect	Britain's Forests and Woodlands
Expand	Britain's Forest Area
Enhance	the Economic Value of Our Forest Resources
Conserve	and Improve Biodiversity, Landscape and Cultural Heritage of our Forests and Woodlands
Develop	Opportunities for Woodland Recreation
Increase	Public Understanding and Community Participation in Forestry

5.2 Restructuring of the Forestry Commission

In the minds of many people, forestry is a profession and a business where life both in the forest and in the office moves at a pace that reflects its long term nature—where the time span between the planting and harvesting of a crop of trees is measured in decades, not seasons. Nothing could be further from the truth in Britain at the latter end of the 20th century.

At the centre the Forestry Commission, as the Government's forestry authority and major producer of timber has undergone complete re-structuring and re-organisation to take advantage of the new electronic data-processing opportunities and mechanisation of production that are open to all businesses and to meet new political requirements (e.g. the separation of the regulation of private and state forestry from the actual management of the forests; the hiving off of Forest Research from a Branch of the Forestry Commission to a Research Agency). New political requirements have included the need to re-organise administration to meet the devolved forestry policies of England, Scotland and Wales that will emerge following the setting up of the Scottish Parliament and Welsh Assembly. All this has caused constant moves of staff, closure of offices, the opening of new offices with much uncertainty and disruption to staff and their families.

It is of course much too soon to judge which of the many changes will be truly beneficial, which will prove to have been merely cosmetic and which will result in a downright disaster.

Suffice to note here that in the 1996–1997 Forestry Commission Annual Report and Accounts [**18**] the Commission records that "The staff reductions were supported by an early retirement scheme which has meant that experienced staff at all levels have been lost." Forests vary enormously in their composition, even in such a small country as Britain. Each forest type requires to be managed sensitively and with a great deal of accumulated local knowledge and experience if the long term objectives are to be achieved. Because the loss of experienced staff can result in major mistakes being repeated, in some European countries, foresters spend the whole of their working lives in one area, often continuing the management practices of their predecessors. This can have its dangers too, with inertia replacing active management to meet changing needs and times.

The Forestry Commissioners themselves have no doubts at all about the correctness of the course on which they are now embarked and which they have charted in their Annual Reports covering the last five years of the 20th century. The main issues are summarised below.

6 Revised Grant Aid for Private Forestry

6.1 More Flexible Grant Aid for Private Forestry—Fine Tuning of Woodland Grant Schemes

In the post-war years, there were within the Forestry Commission many discussions as to whether it was reasonable for the Commission to pay the same level of planting grant throughout Britain, given that, for example, the cost of establishing a pine plantation on the dry heather-clad heaths of north and northeast Scotland, were a fraction of those associated with the establishment of a broadleaved or coniferous crop on the fertile sites of south-west England still bearing a scrubby growth of broadleaves from the previous crop.

As already noted (Chapter 7) a higher level of grant was introduced in 1985 for the establishment of broadleaves—the so-called Broadleaved Planting Grant—with the object of increasing the area of broadleaved planting. This was the first (and a highly successful) attempt at "fine tuning" grant aid for forestry.

Other targeted grants were introduced to stimulate the private sector to meet specific objectives or to carry out schemes which the Government thought were important but which the standard Woodland Grant Scheme levels of grant would not achieve or achieve only at a very slow rate. These schemes described below, were the Woodland Improvement Grant, the Locational Supplement and Challenge Funds.

6.2 Woodland Improvement Grant

In 1995 a Woodland Improvement Grant (WIG) was launched as part of the Woodland Grant Scheme (WGS). The first project ran for 2 years and is helping to provide public access to woodlands. The WIG will now be available to improve the management of poor quality woods and for work to enhance biodiversity [**14**].

Further aspects of work to bring woods back into productive management which will receive funding are:

- uneconomic felling; respacing; rhododendron control; protective work in ancient semi-natural woods for conservation of rare species such as the red squirrel, dormouse, capercaillie, butterflies and bats.

6.3 Locational Supplement (LS)

A supplementary grant to the Woodland Grant Scheme was introduced in April 1995—the Locational Supplement—to provide an additional incentive for the planting of new woodlands in specially targeted areas which will have a high environmental and recreational potential. Examples of areas where the LS is payable (as at December 1999) include:

- Central Scotland
- Thames Chase (north of the River Thames between Dagenham and South Ockenden and Brentwood)
- Red Rose (an area bounded by Manchester, Altrincham, Wigan, Horwich and Bury)
- The Mersey Forest (an area to the north, east and south of Liverpool)
- The Aman Gwendraeth Woodland Initiative in Wales

In areas targeted, the emphasis may vary from planting in existing woodland or on bare land including farm land [**14**].

6.4 Challenge Funds

Challenge Funds represent an entirely new concept for providing grant aid to encourage (a) special new planting and (b) management initiatives in existing woods. The funds are separate from, but complementary to, the Woodland Grant Schemes which are the main vehicles for promoting private forestry in Britain.

When the objectives and terms of a Challenge Fund are announced, landowners and other would-be forest owners are invited to bid for the Fund, stating how in detail they would implement the broad proposals outlined in the challenge. To put

Challenge Funds in perspective — "over 3000 hectares were approved in year 1998–1999 to create new, well-designed and productive forests"; in that year woodland grants provided to the private sector led to the planting of nearly 16 000 hectares of new woodland, of which 60% was broadleaves. The Commission comments: "Whilst it is encouraging to see overall planting levels being maintained, especially given continuing high agricultural land values, there is continued scope to encourage more productive conifer planting." Some of the new Challenge Funds are expected to help in this respect.

It is also fair to say that Challenge Funds have been instrumental in creating woodlands in areas and by people where the Standard Woodland Grant Schemes have been (or been seen as being) unattractive, and so have contributed significantly to the establishment of new woodland in very different parts of the country.

7 Forest Design and Longer-Term Forest Plans

In his foreword to the Forestry Commission Annual Report and Accounts 1995–1996 [**19**] the Chairman draws particular attention to the "redesign of our own (FC) forests which will greatly enhance the rural infrastructure. We are encouraging the same process in the private sector so that forests will look more attractive, be richer in biodiversity and offer wide-ranging recreational activities—from the traditional forest walks to more unusual activities such as open-air concerts".

The Forestry Authority acts as the regulatory body for forestry in Great Britain whether it is carried out by private woodland owners or by Forest Enterprise. By March 1996 the Authority had approved design plans covering 400 000 hectares of harvesting and regeneration submitted to it by private woodland owners and by Forest Enterprise. Local and other authorities were consulted on these plans.

The Forest Design initiative used by Forest Enterprise has been extended to the Private Sector.

In 1997 the Government announced proposals to encourage owners of "larger forests" to work to long-term Forest Plans of some 20 years duration. Most of the post (1939–1945) war expansion of Britain's privately owned forests was driven by the Commission's Dedication Scheme, where owners committed themselves to manage their forests "in perpetuity" according to long-term objectives agreed with the Commission and with a detailed Plan of Operations approved by the Commission for ten years ahead. The original Dedication Scheme was closed in 1972 and replaced by a new (Basis III) Dedication Scheme, which committed an owner for his lifetime and gave his successor in title the right to renew Dedication if he so wished.

In 1988, when the Government of the day (without prior warning) took forestry out of the tax system that had been approved by all governments for many decades and which had hugely stimulated private forestry for so long, Dedication was made more and more unattractive to private owners so as to encourage them to move to one of the new short-term five year "Woodland Grant" schemes. These schemes are much more popular with Government—and the Treasury in particular—because they do not commit Government to long-term financial expenditure on private forestry.

There are a number of matters of detail yet to be clarified, including the flexibility that will be built into the scheme to allow a new owner purchasing or inheriting woods during the 20-year period of an agreed plan to change the plan to suit his particular wishes. The importance of this is that commercial woodlands change hands on average every 10 years.

On the whole, the new 20-year schemes for larger private forests is an important change for the better and represents a compromise between the long-term nature of forestry and a realistic time-span for forward planning in a world where the pace of change is far more rapid than it was half a century ago.

In 1999 the Forestry Commission was able to announce [**16**] that: "Our Forest Plans initiative has advanced considerably during the year. It aims to extend the principles of Forest Design Plans as used by Forest Enterprise to the private sector thereby encouraging a long-term, holistic approach to forest management. It will provide a Plan Preparation Grant to private owners to prepare a coherent 20-year management plan. The Forestry Commission will then be able to approve felling and restocking operations for the first 10 years of the approved Forest Plan and reduce the administractive burden associated with the current mechanisms of regular licensing and five year grant approvals.

A Steering Group was established comprising forest industry representatives and Forestry Commission staff. In May 1998, over 40 interested parties attended a forum which provided guidance to the Steering Group's work. A pilot process was initiated in 16 varied locations throughout Great Britain. The Forestry Commission has paid the costs incurred by the pilot sites, which were designed to provide real cost information to help set the Plan Preparation Grant rate.

A feature of the Forest Plans process is an initial scoping(consultation) exercise where applicants are required to seek the views of other interested parties before a Plan is prepared. All those involved in preparing a Forest Plan place a very high value on this exercise. Invaluable experience has been gained and reported back to the Steering Group.

The Steering Group made its recommendations to the Forestry Commission in March 1999, preparing the way for the operational details of Forest Plans to be developed and the initiative to be formally launched."

8 The UK Forestry Standard Forestry Commission 1998. The Government's Approach to Sustainable Forestry

The UK Forestry Standard described by the Commissioners as the "highlight of the year", was published on 20th January 1998 [**2**]. The Standard was developed after wide public consultation and recognises the great range of uses that are made of woodlands in the UK i.e. Great Britain and Northern Ireland.

The Prime Minister in his Foreword states that the UK Standard is more than a document describing good practice; it explains how the Government will monitor the UK forests and take steps to correct anything found to be going wrong. "The Standard explains how we will achieve this through a combination of scientific study and national surveys. We are committed to publishing results and to continued development of the Standard in the light of experience".

The historical background to the Standard includes the recommendations of the report of the World Commission on Environment and Development (The Brundtland Commission) Our Common Future 1987; and the Rio de Janeiro 1992 Earth Summit commitments to sustainable development—Statement of Forest Principles, Agenda 21, the Biodiversity Convention and a Framework Convention on Climate Change. The Ministerial Conference on the Protection of European Forests, Helsinki 1993, interpreted the Rio principles for European conditions in the Helsinki Guidelines (Appendix I of the UK Standard). In 1994 the UK published Sustainable Forestry—the UK programm [**1**].

The international agreements stimulated world-wide interest in sustainable forestry and European countries have agreed some broad Pan-European Sustainable Forest Management Criteria (PEC) to be applied to forests throughout Europe (see below). These are outlined in the 1998 UK Forestry Standard, which claims that the UK Forestry Standard is "compatible with both the Helsinki Guidelines and the PEC."

Environmental guidelines were introduced into the UK before the Helsinki conference and "directly addressed practical management issues related to major components of the forest ecosystem or human resource." The components represent the basic resources of forestry and the UK Standard illustrates how "stewardship of these is readily matched with both Helsinki Guidelines and PEC."

Sustainable Development is defined as:

- development that meets the needs of the present without compromising the ability of future generations to meet their own needs (Our Common Future 1987; report of the World Commission on Environment and Development).

Sustainable Forest Management means:

- the stewardship and use of our forests and forest lands in a way, and at a rate, that maintains their biodiversity, productivity, regeneration capacity, vitality and their potential to fulfil now, and in the future, relevant ecological, economic and social functions at local, national and global levels and that does not cause damage to other ecosystems (Ministerial Conference on the Protection of European Forests, Helsinki, 1993).

The Pan-European Criteria (PEC) proposed in 1994 to be applied to forests throughout Europe and quoted in the UK Forestry Standard [**2**] are:

- maintenance and appropriate enhancement of forest resources and their contribution to global carbon cycles.
- maintenance of forest ecosystem health and vitality.
- maintenance and encouragement of productive functions of forests (wood and non-wood).
- maintenance, conservation and appropriate enhancement of biological diversity in forest ecosystems.
- maintenance and appropriate enhancement of protective functions in forest management (notably soil and water).
- maintenance of other socio-economic functions and conditions.

The Standard identifies through a descriptive process the forestry practices which are appropriate or inappropriate in particular situations. Sources of information and relevant publications are identified and listed in a comprehensive publications and keyword index.

9 Home Grown Timber Advisory Committee Replaced by Forestry Commission's Advisory Panel[2]

In 1998 the Forestry Commissioners decided to replace the Home Grown Timber Advisory Committee (HGTAC) of 25 members by a smaller 15 member Forestry Commission's Advisory Panel (FCAP). The HGTAC met for the last time in March 1999 and the FCAP met for the first time on 15th July 1999 (Forestry Commission, personal communication) and not in June, September, and December 1998 and March 1999 as reported in Appendix 3 of the Forestry Commission Annual Report 1998–1999 [**16**]; it was the HGTAC that met on those dates (see Appendix 4 of Ref. [**16**]).

10 The Future for Forestry. A Frame-work for Forestry in Great Britain

Advice to the Forestry Commission from the Home Grown Timber Advisory Committee (HGTAC)—a paper presented to the Commissioners in May 1998 [**20**], prior to its demise.

> *Note*: On the "Contents" page the paper states: The Home Grown Timber Advisory Committee is the Forestry Commission's central advisory body set up under the Forestry Act 1967. It advises the Forestry Commissioners on their statutory duties and is concerned with promoting the interest of forestry, the production and supply of timber and with the environmental and social contexts of forestry.

---

[2]The Forestry Commission Advisory Panel is (wrongly) given the acronym (HGTAC) in the 1998–1999 Forestry Commission Annual Report.

But the HGTAC existed long before 1967. Originally set up in 1938, the HGTAC was concerned with matters relating to the supply of home-grown timber during the 1939–1945 war. Post-war, it advised the Forestry Commissioners on their general duties in an informal way, but its role was strengthened following the report of the Departmental Committee on the Marketing of Woodland Produce (the Watson Committee) in 1956 (see Chapter 3).

The Summary and Recommendations of the Future for Forestry are:

Sustainable Management

Sustainable forestry requires multi-purpose management to realise the wide range of benefits that forestry can bring—economic, environmental and recreational.

The production of wood will remain the main source of forest income for most forest owners in the UK. Only 15% of the timber consumed in the UK is home-grown. Wood prices are therefore set in an international market and competitive pressure is becoming intense.

British forests are generally well managed within a well-tried institutional framework. The introduction of the UK Forestry Standard will reinforce the drive for effective management in the forestry industry.

People and the Cultural Setting

The vast majority of people in Britain live in towns. Forests and woodlands, particularly those close to towns, can contribute greatly to the relaxation and recreation in the countryside that people seek.

The HGTAC recommends a target to plant 100 000 hectares of new urban woodland over 20 years.

In some rural areas, the jobs generated by forestry play an important part in sustaining local economies. Although technological advances mean there are fewer jobs in direct forest activities, these losses have been compensated by new opportunities in tourism and conservation. The Common Agricultural Policy may encourage more planting of forests on better land as an alternative to agriculture and to strengthen the rural economy.

Timber Issues

Producing sawn timber is the key to profitability for commercial forestry. Home producers have won the major share of the lower-value sawn-softwood market. Improvements in sawmilling technology have enabled home producers to break into the construction market, where competition from overseas is most severe. Expansion of this break-in is vital to the future growth of the industry. On the HGTAC's recommendation, the Forestry Commission is undertaking a new study to explore opportunities for developing this market.

Growing timber for the construction industry puts a premium on quality and HGTAC recommends that the forestry industry switch its emphasis from quantity to quality. This has implications for silvicultural techniques and for research.

Planting on better land should lead to higher yield classes with improved profitability, which will encourage growers to invest in good silviculture. But that is for the future. The priority for the industry now must be to optimise the use of existing resources.

By 2025 the production potential from existing forests will reach 15 million m^3 a year of softwood. Thereafter domestic supply will dip, with consequences for investment and standards of forest management. To counteract this HGTAC recommends a programme of new conifer planting, initially 11 000 hectares a year, aimed at adding 275 000 hectares to the country's production resources.

Good quality hardwoods can be readily grown in Britain. However, the low rate of investment return due to the slow growth of broadleaved trees discourages some potential growers.

Broadleaved trees (the traditional cover of much of the British Isles) make a major contribution to the landscape and the non-market values that the countryside can provide. The availability of better land provides an important opportunity for broadleaved planting in pure or mixed forests. The HGTAC suggests parity with conifer planting and recommends an initial annual new planting target of 11 000 hectares, with the longer-term aim of adding 275 000 hectares to the country's broadleaved woodland resource. Some 5000 hectares of this annual target will be native woodland to be considered as part of the Biodiversity Action Plan.

Environmental Issues

The post-war era has left a legacy of uniform forestry, particularly on uplands. Forestry guidelines and the UK Forestry Standard point the way to forest management that will improve landscape values and biodiversity while sustaining profitability. Concern for the environment and landscape and a wider appreciation of the non-timber attributes of forests have altered society's priorities and encouraged a more varied multiple-use philosophy in forestry policy.

The HGTAC endorses the value of woodland habitats in the Biodiversity Action Plan. It supports the emerging targets for improving the condition of semi-natural woodland, including a target for initiating up to 5000 hectares a year of new native woods.

Author's Note: The paper, The Future For Forestry—which is sub-titled A Framework For Forestry In Great Britain—is, on the whole, well balanced. It draws attention to what it perceives as weaknesses in British forestry and forest industries policies and practices, namely a tendency to go for quantity at the expense of quality, and in the past insufficient emphasis on the non-wood benefits of the forest such as amenity and nature conservation.

On the other hand, it may be said that the report undervalues some of the past achievements of British forest policy. By way of example the report states that:

> British forests are generally well managed within an effective institutional framework, but there needs to be more emphasis on the *long-term*

management (author's italics) of woods, whether for timber production or other benefits.

The HGTAC makes no mention of the main post-war (1939–1945) scheme for the rehabilitation and great expansion of private woodlands—the Dedication Schemes (Basis I and II) which bound owners entering the scheme and receiving public subsidies either directly by way of grants or indirectly through the tax system, to manage and maintain their woodlands as woodlands in perpetuity, by covenanting that in perpetuity their forest land would not be used for purposes other than forestry.

The Dedication Scheme Basis III was less demanding on owners; but they entered into the scheme for life with the option of their heirs and successors continuing the scheme.

As noted in Chapter 7, the sudden abandonment of these long-term schemes in 1988, was a Government decision imposed on the Commissioners. Successor schemes have been limited to 5 years—which is as short term as you can get in forestry—until 1998 when 20-year plans were envisaged by the Forestry Commission for some owners of larger woodland estates.

11 Environmental Assessment (Forestry) Regulations 1988

In September 1988 the requirement that forestry projects receiving grant aid from the Forestry Commission and that were likely to have a significant effect on the environment needed an environmental assessment, was extended to cover all forestry projects likely to have a significant environmental impact [**16**].

These new regulations include:

- Creating new woods and forests (including planting Christmas trees and short rotation coppice and creating woodland by natural regeneration).
- Constructing forestry roads.
- Quarrying to obtain material for forestry roads.

At first the regulations were applied in only a few cases, but by the end of the century large numbers of grant-aid applications were affected (John Everard, Chairman Continuous Cover Forestry Group, personal communication).

12 Plant Health

The Forestry Commission continued to inspect at ports imports of sawn softwoods, wooden packaging material and other assortments (e.g. pulpwood from home forests) as part of a major campaign to detect and destroy insects potentially dangerous to Britain's forests. In the two years 1997–1998 and 1998–1999, more than 13 000 inspections were carried out at ports on imported sawn softwoods alone [**16**].

In April 1997 fees for inspection were introduced; powers have been taken to recover costs when supervising the treatment of wooden packing material found to present a risk to plant health.

13 International Forestry Policies and Initiatives

In the period 1995–1999 the Forestry Commission participated actively in the many international bodies now promoting international collaboration in forestry. These include:

- The Forestry Department of the Food and Agriculture Organisation (FAO) and its Committee on Forestry (COFO). The FAO (which is part of the United Nations) has been the leading international forestry body since the end of the 1939–1945 War, and in its early days owed much to the leadership of the Forestry Commission delegates. In March 1999 a meeting of Forestry Ministers convened by FAO adopted a resolution emphasising the critical importance of forests and recent progress in supporting and promoting sustainable forest management, the Rome Declaration on Forestry—see below.
- The European Union, which *inter alia* is studying the role of forestry as an alternative form of land use to farming in the context of the reform of the Common Agricultural Policy (CAP).
- The Intergovernmental Forum on Forests (IFF) established in 1995 under the auspices of the United Nations Commission on Sustainable Development (CSD)
- The United Nations General Assembly Special Session (UNGASS), where forestry was a key issue and featured in most of the national statements delivered by Heads of State or Government. The (UK) Prime Minister made specific reference to the Forestry Commission's commitment to sustainable forest management and the development of the UK Forestry Standard.
- The World Forestry Congress held in Turkey in 1997 where the theme was "Forestry for Sustainable Development Towards the 21st Century" (the Congress was attended by more than 4000 forestry experts from some 150 countries).
- A G8 group of forest experts was convened in 1997 to prepare a programme of practical action on forests to be supported by the G8. The Forestry Commission hosted a follow-up meeting in preparation for the G8 Summit in the UK in 1998. A report on the implementation of the G8 Action Programme for Forests will be made to the G8 Summit in the year 2000.
- European Union Standing Forestry Committee—where the Forestry Commission used its Presidency role in 1998 to chair discussions on the role of forestry in the reform of the Common Agricultural Policy (CAP).

13.1 International Action Programmes and Declarations on Forests and Forestry

Of the many pages of text issued by international agencies and conferences on the subject of forests and forestry during the last 5 years, the following extracts from reports by the G8 countries (1998) and FAO (1999) summarise the key concerns that have been expressed in the closing years of the 20th century.

13.2 G8 Action Programme on Forests; Extracts from G8 Birmingham (England) Summit 9th May 1998

Source:
International Policy, Forestry Commission, Edinburgh E12 7AT.
e-mail: libbyjones@forestry.gov.gsi.uk [**3**]

Introduction

1. The following elaborates by a Programme which reflects a political commitment and aims to complement the extensive range of actions currently being taken by the international community and various regional and international processes and to strengthen some activities G8 members have identified as issues of particular importance. The Programme focuses on domestic actions in the G8 member countries and areas where they can make unique contributions through their bilateral assistance programmes and through their support for intergovernmental processes. G8 members intend to follow up this action programme individually and/or co-operatively and to review and report on progress as appropriate to G8 summits.

Monitoring and Assessment

2. The G8 members participate in international processes within which national level criteria and indicators for sustainable forest management have been developed. These criteria and indicators are tools for monitoring and assessing national trends across land ownerships in forest conditions and forest management. As such, they provide a common framework for describing, monitoring and assessing, over time, progress towards sustainable forest management. The link between national level criteria and indicators and the Food and Agriculture Organisation's ongoing global forest resources assessment programme is also important in providing consistent, reliable and compatible forest data on a global basis.
3. The G8 members will:

- monitor and assess the state of their own forests using agreed national level criteria and indicators for sustainable forest management and make the results, including areas where additional information is needed, available to interested parties;

- drawing on national level assessments, provide information and data to the Food and Agriculture Organisation's global forest resource assessments and particularly Forest Resource Assessment 2000;
- work with partner countries to build national capacity to: participate in regional criteria and indicator processes; develop and apply agreed criteria and indicators to monitor and assess the state of their own forests; develop national forest inventory and monitoring systems which take account of these criteria and indicators; improve scientific underpinning of the economic, social and environmental indicators of sustainable forest management;
- improve access to remote sensing data and geographic information processing technologies, including geographical information systems (GIS) and global positioning systems (GPS), to groups and organisations with an interest;
- exchange information and experience with partner countries on monitoring and responding to large scale disasters affecting forest ecosystems, such as forest fires.

National Forest Programmes: Extracts

National forest programmes to promote sustainable forest management will contribute to national strategies for sustainable development which the United Nations General Assembly Special Session called on all countries to formulate by 2002.

The G8 members will share their experience and focus technical and financial assistance on countries which give priority to sustainable forest management.

They will co-operate with and support the similar objectives of other international bodies such as the International Tropical Timber Organisation (ITTO) and the International Forum on Forests (IFF).

Forests contain 70% of the earth's terrestrial biodiversity. A better understanding of protected area (of forests) management is needed.

G8 members will encourage and help the private sector (including forest-based industry) in the development of voluntary codes of (good) conduct and sustainable management regimes. They will help to develop practical counter measures to illegal logging, in co-operation with, for example, ITTO and IFF.

13.3 FAO—Rome Declaration on Forestry, 9th March 1999 Signed and Adopted by 127 Countries [4]

We, the Ministers responsible for forests or their representatives who met at FAO Headquarters in Rome, Italy on 8th and 9th March 1999 at the second Ministerial meeting on Forestry under the sponsorship of the FAO, to consider "Sustainability Issues in Forestry, the National and International Challenges."

emphasising the critical importance of forests to the welfare, livelihood and food security of all people and future generations and to the life support system of the entire planet, as set forth in the 1995 Rome Statement on Forestry and in the

1996 World Food Summit Plan of Action,

deeply concerned with the important challenges associated with forest loss and degradation in many regions and **stressing** the need to maintain the integrity of forests and ecosystems by promoting sustainable forest management worldwide,

noting that this Ministerial Meeting was convened by the Director-General of FAO to highlight the critical need to make national and international progress toward the sustainable management of the world's forests, and to consider international instruments to support sustainable forest management, global action to address forest fires, and the Strategic Framework for FAO 2000–2015,

welcoming the progress made to date on sustainable forest management, including FAO's role in the assessment of forest resources, and the significant achievements at national, regional and international levels, including the development and implementation of criteria and indicators for sustainable forest management and of national forest programmes,

noting that the Intergovernmental Forum on Forests (IFF), which was established under the auspices of the United Nations Commission on Sustainable Development with a time-limited mandate, is considering issues related to sustainable forest management, including international arrangements and mechanisms,

also noting a range of options for international arrangements and mechanisms to support sustainable forest management, including voluntary, incentive-based approaches; private sector initiatives; regional agreements and initiatives; and global legally binding and non-legally binding instruments,

further noting that a number of international arrangements and mechanisms already exist which address and have implications for the environmental, social and economic aspects of sustainable forest management, and that the need for and nature of future international arrangements and mechanisms should be considered in more depth by the UN Commission on Sustainable Development at its eighth session in the year 2000 as a matter of priority,

recognise that this Ministerial Meeting provides an opportunity to exchange views on international arrangements and mechanisms and provide high level political support to the ongoing discussions at the Intergovernmental Forum on Forests,

call on FAO to facilitate and give support to national, regional and international processes related to forests, especially enhancing the implementation of national forest programmes and criteria and indicators for sustainable forest management,

noting that the causes of forest fires are many and complex and **recognising** the need to harness efforts to prevent forest fires as well as to address the multiple causes and consequences of fires around the globe,

welcome the meeting on Public Policies Affecting Forest Fires hosted by FAO in October 1998 and encourage FAO to take action to implement the recommendations directed to it,

call on FAO and other international organisations, donor agencies and interested countries to work together to address the underlying causes of forest fires, to improve the co-ordination of their efforts to prevent and combat forest fires and to rehabilitate affected areas with a view to providing assistance requested by governments,

noting the draft Strategic Framework for FAO 2000–2015 that will be considered by the FAO Conference in November 1999 highlights the critical need for countries to implement integrated approaches to sustainable development, including combating desertification and drought, which recognise the importance of the sustainable management of forests and their ecosystems in achieving sustainable agriculture and food security,

also noting that forests can be adversely affected by policies outside the forest sector, including agriculture and trade policies and **recognise** the need for integrated land use and mutually supportive trade and environment policies in support of sustainable forest management,

underscore the importance of inter-disciplinary and cross-sectoral approaches to forest management, participatory decision making in development taking into account gender balance, and increased co-operation, including through partnerships between the public and private sectors to achieve sustainable forest management and sustainable development,

encourage national, regional and international efforts to increase public awareness of the importance of forests and strengthen education, research, extension and the dissemination of knowledge and information to improve forest management,

call on all interested parties to give greater priority to sustainable forest management,

reaffirm the proposals for action agreed by the Intergovernmental Panel on Forests regarding international co-operation in financial assistance and technology transfer;

recognise the need to build capacity for sustainable forest management in public and private sectors at local, national and international levels; and **urge** FAO to allocate adequate financial resources to forestry,

welcome the recommendations of the Fourteenth Session of the Committee on Forestry (COFO) and encourage their endorsement by the FAO Council with a view to facilitating their early implementation.

We pledge our political will to bear on improving forest management in our respective countries and to promote effective international co-operation to achieve sustainable forest management worldwide.

We further pledge

- to work together towards a constructive and forward looking outcome on future arrangements for the global forest policy dialogue at the eighth session of the United Nations Commission on Sustainable Development;
- to better co-ordinate and strengthen our efforts to prevent, manage, monitor and suppress forest fires, especially in anticipation of the next E1 Niño/La Niña events and, in the longer term, to address the underlying causes of forest fires;
- to work closely with our counterparts in other ministries in our countries to promote cross-sectoral policies and activities that support sustainable forest management.

14 Conserving Biodiversity

The Forestry Commission in its Forestry Authority role and through Forest Enterprise continues to make a major contribution to biodiversity conservation throughout Britain. In 1998–1999, it established a Forestry Commission Biodiversity Group to co-ordinate programmes and policies; launched an Ecological Site Classification Scheme; and issued guidance on soil conservation and forest design planning.

The final report was received by the Commissioners from the Native Woodland Advisory Panel for Scotland that it had appointed earlier.

15 Forest Recreation

The Forestry Commission, through Forest Enterprise, is the largest single provider of outdoor recreation in Britain, with around 50 million visitors a year.

Private woodland owners receiving grant aid for forestry are encouraged to give public access to their woods and forests where this is feasible; the Walkers Welcome signposting now actively invites people to enjoy nearly 70 000 hectares of woodland. In some instances for woods which are managed under a Woodland Grant Scheme, funds are available to open areas to the public or to enhance areas already open to the public.

16 Hunting and Deer Management

The Forestry Commission's position on Hunting was in 1999 that all Fox hunting in the Forestry Commission's public forests was licensed under a revised Master Agreement and concluded that year with the Masters of Foxhounds Association.

No licenses were issued for deer hunts.

An agreement was reached with the Masters of the Draghounds and Bloodhounds to allow drag hunting—a relatively new activity—to take place on Commission land.

Deer, which are in the main woodland creatures, can and do cause great damage to their habitat if numbers are not controlled. Red deer, for example, are one of the main enemies of the remnants of the native Scots pine woodlands of Scotland because they destroy the natural regeneration on which the long-term survival of the forest depends—the replacement of the ancient veterans by a gradual succession of younger trees of all age classes down to seedlings. In addition to the native deer—red and roe—there are increasing numbers of introduced species which have escaped from the "parks" where they were kept by wealthy landowners as pets.

The problem of over-population of deer is a European wide one and so far has not been solved.

In Scotland the control of Red Deer was recognised as a necessary element of good land use in 1959 by the establishment under Act of Parliament of the Red Deer Commission. *Inter alia* it was hoped that it would help resolve the potential conflict between the sporting estates dependent financially to a large extent on deer stalking and forestry interests. This conflict incidentally, might well be present on the same estate with gamekeepers and foresters each fighting for their own interests.

In 1996, Parliament passed an Act bringing all species of deer under the Commission—now titled the Deer Commission (for Scotland).

England and Wales have relied and continue to rely on an agreement between local and regional interests on such deer control (if any) as might be thought to be necessary. Government has encouraged and strengthened the formation of these voluntary bodies in recent years and given them support and some modest financial support to help in running costs.

17 Public Participation and Communications with the Public

For many years the role of society in formulating a country's forestry policy has been discussed at international forestry meetings, notably those of the Forestry Division of FAO, including, for Europe, the European Forestry Commission. The Commonwealth Forestry Conferences have also in recent years devoted considerable attention to the subject (Chapter 12).

The current situation for Britain is well summed up in the Forestry Commission 1998–1999 Annual Report [**16**] under "Partnership Working".

"Over the last few years, forestry policy has evolved to embrace the role that woods and forests can play in meeting society's wider needs, for example helping to overcome rural disadvantage and social exclusion, and helping to protect and enhance habitats. One consequence has been an increase in the number and range of different interests who have come together to work on achieving a common goal. These partnerships have demonstrated how different expertise, skills, knowledge and resources can be combined to deliver a common objective or goal. Working through partnership is one of the Government's underlying principles for implementing its policies and this approach is at the heart of its Forestry Strategy for England.

A wide range of partnership based initiatives have already been established including, for example, the 12 Community Forests in England, the Forest Education Initiative, Highland Birchwoods and the Deer Initiative. Forest Enterprise is involved in a myriad of partnerships the length and breadth of the country. Often these partnerships are able to increase the total resources available to implement their objectives, by unlocking new resources such as lottery and European funding and through sponsorship from public and private bodies.

In order to strengthen our ability to support partnership working we are ensuring that our staff are trained in best practice in this area and we are adjusting our funding mechanisms. We have established a Forestry Partnership Programme to ensure that our input to shared projects and initiatives is fully effective and to target our resources more effectively. The Partnership Programme is complementary to our funding through the Woodland Grant Scheme."

It has often been said by foresters that they are not good at communicating the benefits of forestry to the general public. In recognition of this, the Forestry Commission has taken professional advice from specialist consultants in the communications and public relations field and started a training and implementation programme for all its staff.

Forest Education Initiatives (FEI) have been strengthened; these connect forestry and wood with local schools and other educational establishments and help in the teaching of literacy and numeracy. Links between schools and colleges and their local Forestry Commission staff and woods go back to the early days of the Commission. In addition to providing opportunities to create and maintain small woodland plots and interest children in the forest life around them, days in the woods measuring tree sizes and volumes and reporting on what wildlife they encountered were found by imaginative teachers to be fruitful ways of instilling some basic geometry and basic English into pupils bored with the same subjects taught formally in the classroom. And of course such days helped instil a respect for their woodland environment around them.

Agricultural shows have also been used by the Forestry Commission, private woodland owners and timber merchants to put across the benefits of forestry. An innovative approach was the use of a puppet theatre at Scotland's Royal Highland Show, which "drew in huge crowds to hear about the importance of forests to wildlife" [**16**].

18 The Progress of Private Forestry 1995–1999

The area planted in Great Britain by Private Woodland Owners for which grants were paid in the five years 1995–1999 are given below. The table includes planting grant-aided under the Woodland Grant Scheme (WGS) in association with the Farm Woodland and Farm Woodland Premium Schemes administered by the Agriculture Departments.

Table 8.5 Area planted by private woodland owners in Great Britain for which grants were paid in hectares

Year	New planting	Restocking	Total
1995	18 128	6169	24 297
1996	14 975	5761	20 736
1997	15 971	6533	22 504
1998	15 915	5525	21 437
1999	15 914	5486	21 400

Source: Forestry Commission Annual Report and Accounts 1998–1999.

The total of these 5 years of 110 000 ha has been included in the summary in Section 3.2 of this chapter to complete the record of this century.

The area of private woodlands managed under a Dedication Scheme had fallen from 246 392 hectares in 1994 to 147 264 hectares in 1999 and under a Woodland Grant Scheme (including WGS associated with the Farm Woodland Premium Scheme) had increased from 575 446 hectares in 1994 to 684 300 hectares in 1999, giving a total area of 831 564 hectares of private woodland managed under a Plan of Operations approved by the Forestry Commission in 1999 (an increase of 9726 hectares over the 1994 total).

19 Planting Rates over the 20th Century

Figs. 8.1 and 8.2 summarise the previous tables for planting over the century; Fig. 8.1 for the Forestry Commission, Fig. 8.2 for Private Woodlands. For the Commission, entirely dependent on Government support, planting trebled the pre-war rate for some 30 years, but has since been cut back to simply re-stocking what is felled.

For the private sector, Government grant aid and taxation relief account for the overall rate of about 15 000 ha per year in the second half of the century, more than four times that achieved earlier. The major hiccup from 1976 was due to the ending of Dedication Basis I and II, and later removal of tax benefits combined with the failure of Dedication Basis III to take off. Clearly the success of the Forest and Woodland Grant Schems have, to a large extent, restored the overall level of planting. But one wonders whether the shift of emphasis to broadleaves with the loss of the Forestry Commission's contribution will not have dire consequences for wood production mid-21st century?

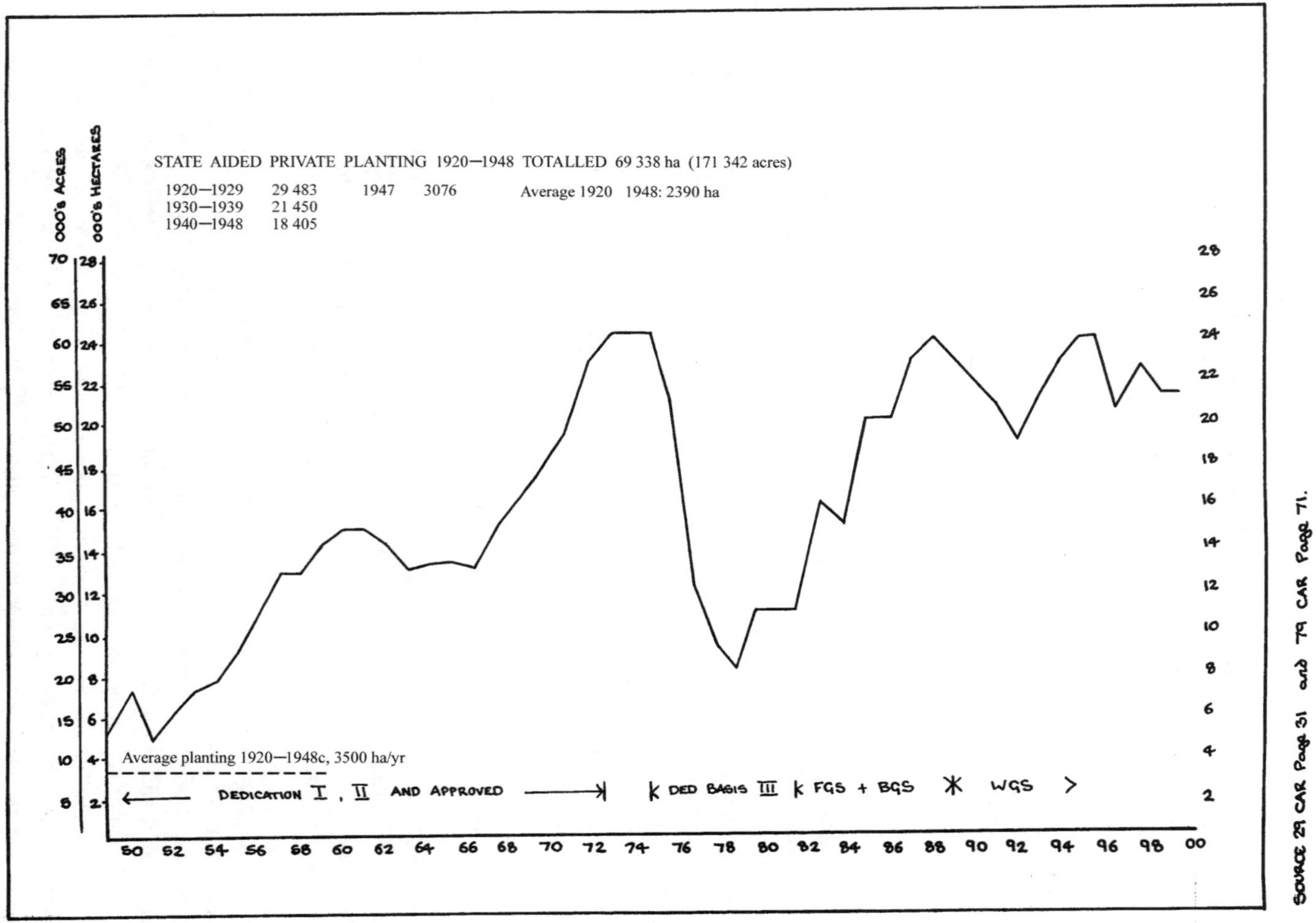

Figure 8.1. Forestry Commission Planting—Aforestation and restocking 1949–1999

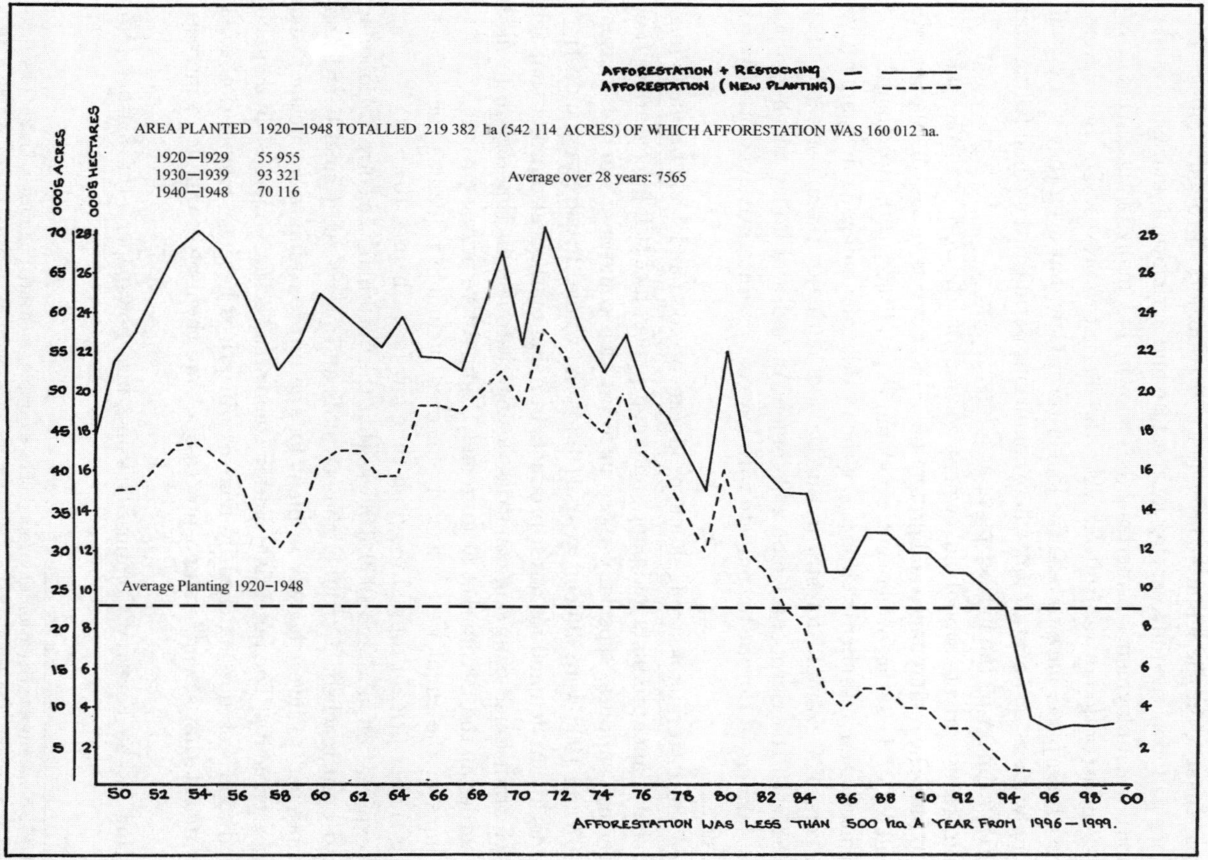

Figure 8.2. Private Woodlands Planting—Aforestation and restocking 1949–1999

20 The Longer-Term Outlook for British Wood and Wood Products

One important direct effect of the low prices of coniferous timber (see below) is that many owners postponed thinning and clear felling with the resultant loss of jobs for people engaged in the felling, conversion and transport of round timber. The closure of small local sawmills serving local estate and agricultural markets can also create local unemployment. Experience has shown that many of those that leave the forest industry do not return to it when the market improves. The big players, like Forest Enterprise, have to take the longer-term view, but even they will lay off harvesting workers and contractors where conditions permit rather than felling and transporting timber to the market place at a loss.

Again, many private woodland owners depend on the revenue from sales of timber to fund the maintenance and protection of their woods, so loss of sales income will result in job losses as owners cut woodland expenditure to the minimum.

The international timber trade is one of the world's remaining highly competitive free markets. For example, forestry in Europe does not enjoy the massive subsidies associated with the European Union's Common Agricultural Policy which enables high cost farmers in Europe to compete with low cost producers in some other parts of the world.

When the £ sterling is strong, home producers of wood and wood products (e.g. sawn timber, panel products and paper pulp and paper) face stiff price competition from overseas supplies, especially those in forest rich countries. For this reason, both 1997 and 1998 were difficult years [**16**] for the whole British forest industry[3]

The decline in demand for forest products in Asian markets also meant that the forest-rich traditional exporting countries, losing customers in this region, pushed all the harder at the UK market at extremely competitive prices.

All links in the chain were adversely affected including of course the grower. Standing timber prices dropped to an all time low (see Chapter 10).

The expectation in March 1999 [**16**] was that the bottom of the current economic cycle had been reached. Certainly the sawmills and processors (panel board and pulp) continued to invest heavily in plant for the future to improve efficiency and to increase capacity. The potential—and in some areas actual—increase in demand for roundwood was not matched by increases in prices because competition from imports meant that sawmillers and processors were themselves unable to increase prices.

Writing in the Forestry Commission's in-house newspaper—the Slasher (No.

[3]In the 1970s a similar situation developed. For example, measured against a basket of wholesale prices of 15 items, the value of the £ sterling rose (in real terms) from a base of 100 in 1973 to 137 in 1980; in this period the American and Canadian dollars fell to under 100. No wonder then that the UK forest products manufacturers found it extremely difficult to compete on price with importers [**21**].

389, December 1999)—the Director General comments on the drastic price drop in the British wood and wood products markets in 1998 and 1999 (as evidenced by the fall in British standing coniferous timber prices—see Chapter 10).

He cites two global factors and one domestic factor which have added to the downward pressure on the wood and wood products market in Britain, over and above the very strong British pound (which, as noted above, has historically lowered prices, for both British growers and all the British forest products industries).

The global factors are:

- the Asian economic meltdown (the Asian developing economies are major importers of forest products)
- the breakdown of the Soviet block which released large quantities of high quality timber from the Baltic forests

The domestic factor is:

- incentives to recycle fibre and waste wood

Reviewing these factors at the end of 1999 he says that:

- the Asian economy is recovering with the Asian bank revising growth forecasts for the "tiger economies" upwards of 5% per annum
- international pulp prices are increasing
- panel products and sawmilling interests are anticipating a building expansion based on timber framed buildings in Eastern Europe
- there are signs that the market for the (necessarily limited supply of) recyclable material is tightening
- British timber importers are forecasting a rise in the price for some coniferous sawnwood in the year 2000.

Noting that the major (1998–1999) downturn in prices is the third that he has experienced in 22 years international forest products trading, with a complete recovery always following, he backs a moderate recovery in the Commission's timber income but suggests that if the European economy strengthens relative to the pound sterling, this prediction could be over cautious.

20.1 Market Development for Softwoods (Coniferous Timber)

In 1999 the first major UK market development study for softwoods for 15 years was completed. It showed that over the next 20 years softwood supply will rise by some 6 million m^3 of which 5 million will be suitable for sawing. The greater part of the increase will be in Scotland. About 67% will come from the fragmented private sector. An additional supply of 2 to 3 million m^3 of wood residues of one sort and another will need to be taken up over the same period.

The Forestry Commission and development agencies have a key part to play in making this resource attractive to wood processing companies willing to invest in the highly capital-intensive plant needed to convert the roundwood and residues into sawn lumber, panel products and paper pulp.

20.2 Market Promotion of Wood

In European Timber Trends and Prospects to the year 2000 and beyond [**17**] attention was drawn to the fact that all the individuals, companies and organisations concerned with the growing, production and marketing of wood and wood products to the final consumer (the so-called "wood chain" or filiere bois) had a common interest in promoting the use of wood and wood products (especially in face of competition from other materials). It deplored the tendency of each participant in this chain to maximise its own financial returns, sometimes at the expense of other participants in the chain. It called on participants to improve this situation by aiming to increase consumption—and thereby the financial returns for all.

These sentiments were again expressed—albeit in different words—in the Forestry Commission's Annual Report for 1998–1999 [**16**], against the background that the UK market for wood in recent years has not been expanding due to the (low) level of building construction and partly due to the success of competing materials such as concrete, steel and plastic. The Report notes that the Forestry Commission "has been playing an important role in facilitating the discussions between UK-based companies and associations representing companies in exporting countries".

References

1. *Sustainable Forestry—The UK Programme.* Cm 2429. HMSO, London 1994.
2. *The UK Forestry Standard.* Forestry Commission, Edinburgh, 1998.
3. *Forestry Commission G8 Action Programme on Forests 9th May 1998.* International Policy Division, Edinburgh.
4. *Rome Declaration on Forestry 9th March 1999.* Second Ministerial Meeting on Sustainability Issues in Forestry. The National and International Challenges. Forestry Commission International Policy Division, Edinburgh.
5. Forestry Commission. 71st Annual Report and Accounts 1990–1991.
6. Forestry Commission. *Forestry Policy for Great Britain.* 1991.
7. *The Forestry Act 1991.*
8. 71st CAR.
9. *The Crofter Forestry (Scotland) Act 1991.*
10. 36th CAR.
11. 47th CAR.
12. 72nd CAR.
13. 73rd CAR.
14. *Forestry Commission Annual Report and Accounts 1994–1995.* HMSO, London.
15. *Forestry Commission Annual Report and Accounts 1993–1994.* HMSO, London.
16. *Forestry Commission Annual Report and Accounts 1998–1999.*

17. *European Timber Trends and Prospects to the year 2000 and beyond Volume 1*, United Nations, New York, 1986.
18. *Forestry Commission Annual Report and Accounts 1996–1997*
19. *Forestry Commission Annual Report and Accounts 1995–1996.*
20. Forestry Commission. *The Future for Forestry. A framework for forestry in Great Britain.* Advice to the Forestry Commission from the Home Grown Timber Advisory Committee, May 1998.
21. E.G. Richards, Allgemeine Forst Zeitschrift No. 30, July 1982.

Chapter 9

Crofters, Crofting and Forestry

1 What is a Croft? who is a Crofter?

The Crofters Commission, which was set up in 1955 to reorganise, develop and regulate crofting in the crofting areas of Scotland (the former counties of Shetland, Orkney, Caithness, Sutherland, Ross and Cromarty, Inverness and Argyll) makes the important point that a croft is not—as many people think—the house in which a crofter lives; it is the land which he occupies.

A crofter is the tenant of a croft who pays rent to the landlord of the croft. A tenant may buy his croft or any part of it, but under the Crofting Acts in so doing he ceases to be a tenant. Because there is no tenant the croft land becomes in law technically vacant. The Crofters Commission have the power to require that croft land which has become vacant be let to a crofting tenant. In practical terms this does not create a problem for the new owner (the former crofter) so long as he or a member of his family continues to occupy and work the acquired land for agricultural purposes [**1**].

There is provision in the legislation for the owner of a croft with a vacant tenancy to apply to the Crofters Commission for a direction that it shall cease to be a croft. In considering the application, the Commission have to have regard to the general interest of the crofting community and the general demand for crofting land in the area. In passing, it should be noted that, according to the Crofters Commission, there is a very "high demand for crofts, not least from the existing crofting population, which far exceeds the numbers which become available."

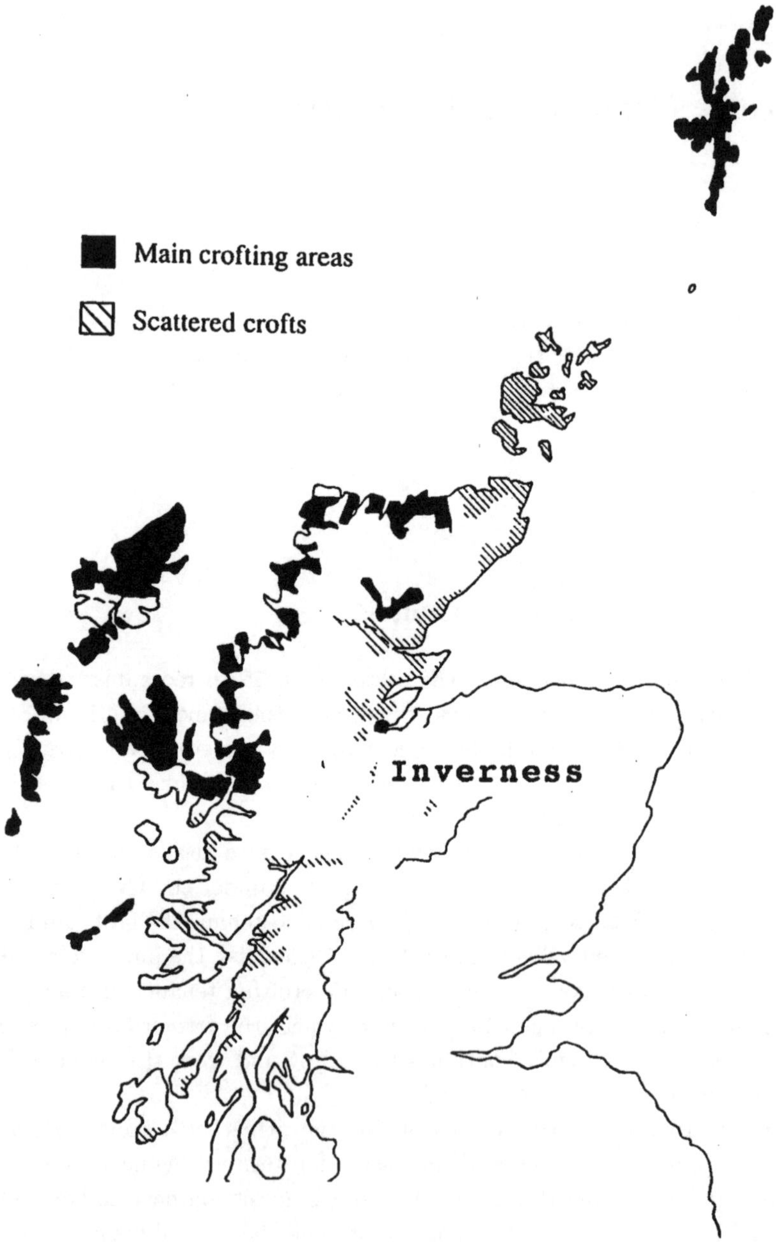

Figure 9.1. Crofting lands

One other point of interest where crofting differs from ordinary tenanted farms is that it is the tenant, not the landlord, who provides the house, steading and other permanent equipment on the croft. A very few crofts have been equipped by the landlord.

In 1982, there were some 17 700 registered crofts in the crofting counties of which 2580 were owner-occupied. The map shows that the main concentrations of crofts are in:

Western Isles	6000
Skye and inner Hebridean Islands	1840
Shetland	2700
North and West periphery of the mainland	2300

Crofts vary in extent from 0.5 to 40 hectares, but the average-sized croft is about 4 hectares of "in bye" land with a share in a large area of common pasture for stock grazing. In the Western Isles and in Shetland which are the main crofting centres, agriculture is based mainly on rearing cattle and sheep which are sold for fattening to lowland farmers. On the east coast with its more favourable climate and soils, there is a little dairy farming, while potatoes, vegetables and soft fruit are grown commercially.

Few crofts—less than 5%—provide full-time employment. Most crofters have ancillary occupations to augment their income from the croft. These include the provision of tourist accommodation and work in the service industries associated with tourism, fishing, fish farming and to a small extent, employment in forestry. Some crofters can be described as hobby crofters.

The absence of crofters woodlands in the crofting counties and the income from them is in stark contrast to the situation regarding small owner-occupied farms in many European and the Nordic countries (Finland, Norway, Sweden). The situation in Norway well illustrates this (Norwegian Forestry Society 1957).

More than 70% of the productive forest area belongs to individuals, a major part of this area being utilised in combination with farming. The total area of these farm forests constitutes some 48% of the total productive forest area of Norway and usually contain the best forest land.

Many of these farm forests are small; thus some 95 000 individual forest properties (17% of the forest area) were under 50 hectares in size, the range being from under 2.5 hectares to 50 hectares. Historically, the forest has provided winter work, especially when the ground has been covered in snow and a cash income. When farm prices have been low, it is recorded that "it was the forest that saved the farm".

In Scotland, crofting land has seldom included woodland and crofters have not been very active in planting trees on their land. The reasons for this are discussed below.

2 Crofters and Historic Barriers to and Lack of Incentives for Tree Planting

The first and over-riding constraints on crofters planting trees have been the legal ones—crofting has been about agriculture and historically crofting legislation has been directed to this end.

Such grant-aid schemes as there were in the inter-war years (1919–1939) to encourage private forestry were in the main directed towards traditional forestry estates and were not aimed at encouraging new small-scale afforestation by individuals. Also, there were constraints on the claiming of Forestry Commission grants for afforestation by crofting tenants.

When—post-1946—the creation of a standing reserve of timber became an even higher national priority, the drive in the crofting counties as elsewhere in Scotland was for large-scale acquisition of plantable land by the Forestry Commission and the re-planting of the private estate woodlands—often felled compulsorily—which had been sacrificed for the war effort.

An important disincentive to crofter forestry that was not addressed until the 1991 Crofter Forestry (Scotland) Bill, was that any trees planted by tenants once they had grown beyond a certain size, were deemed to be the property of the landlord.

The second major factor—and in some locations some would put it first equal with the legal constraints—was the difficulty of establishing new woodlands in the windswept and infertile conditions of climate and soils where much of the crofting land is located.

Additionally, the grants available for the establishment of new plantations were—in the view of many Forestry Commission staff in North Scotland— totally inadequate to compensate crofters for the risks and uncertainties involved in establishing woodlands in many of the crofting regions and especially on the coast.

Various trial plantations and plots were however created, due in no small part to the enthusiasm of Forestry Commission staff keen to try to find out if and how small-scale forests could be created in the crofting counties.

Writing on the potential for tree growth and woodland creation in the Western Isles, Sharpe and Jacyna [**2**] make a number of points which have wider application within the crofting counties. Work on the shelter provided by belts of trees in the Northern and Western Isles continues (C.P. Quine and A.L. Sharpe of the Forestry Commission Research Agency, Newton, Elgin) and assessments of the earlier experiments on choice of species and growth rates continues [**3**].

The broad picture which emerges from all this work can be summarised as follows:

- 18th and 19th century attempts by landowners to grow trees usually ended in failure, but even in the Western Isles (one of the windiest regions in Europe) there are numerous examples of natural woodlands growing in favourable locations.

- More recently, experimental plots established by the Forestry Commission together with information from private plantings and studies of the natural woodlands have demonstrated that quite respectable growth rates can be achieved.
- The major benefits will be for shelter for homes and livestock and the creation of wildlife habitats
- The species that can be successfully grown include not only coniferous pioneer species such as Sitka spruce and Lodgepole pine of Alaskan origin, but a range of broadleaves that can be used to improve diversity and enhance the nature conservation and amenity values of the woodlands.
- Some woodlands will produce commercially utilisable trees.

An important objective of this ongoing research is to find the most feasible methods of creating mixed woodlands—for example, should the pioneer conifers be planted first and then opened up by thinning to allow the introduction of the broadleaves and the native conifer Scots pine, or does this technique involve too great a risk from windblow—suggesting that pioneer and successor species be planted together and the very slow early growth of the successor species be accepted.

3 The 1991 Crofter Forestry (Scotland) Act Heralds New Positive Possibilities for Crofter Forestry

The Crofter Forestry (Scotland) Act received Royal Assent in summer 1991 and the provisions of the act were implemented in April 1992.

The Act empowers grazing committees to use common grazings for forestry and makes them eligible for grants for forestry and farm woodlands. Planting trees on common grazings will not constitute permanent improvements for the purpose of crofting legislation.

The bill takes account of outstanding concerns of the Scottish Landowners Federation and Scottish Crofters Union. Any crofter forestry development requires the consent of both the landlord, to safeguard his interests and the Crofters Commission to consider the impact on the stock carrying capacity of any common grazing and to ensure the view of each member of the crofting township is taken into account. There is a presumption that a Landlord's consent will not be unreasonably withheld.

At this time, the Forestry Authority's Woodland Grant Scheme was expanded to take account of a broader range of objectives and a commitment to timber production was no longer obligatory.

This increased the potential for grant aid to crofter forestry.

The farm Woodland Premium Scheme could provide additional revenue, although limited to 100 hectares per common grazing.

The Act made crofter forestry possible and the grant support available made it attractive to crofters.

Table 9.1 Grants paid to crofting townships for planting on common grazings (by financial year, 1st April to 31st March)

Forestry Authority data			
Year	£	Year	£
92/93	NIL	97/98	250 950
93/94	100 872	98/99	343 132
94/95	89 494	99/00	296 655
95/96	226 749	00/01	1 259 595
96/97	262 471		

Table 9.2 Township Scheme Areas applied for on common grazings

Forestry Authority data	
Financial year (1st April–31st March)	Area (ha)
92/93	None
93/94	408.7
94/95	408.6
95/96	861.6
96/97	583.4
97/98	340.4
98/99	458.7
99/00	511.1
00/01	1976.6

Bearing in mind that there was little crofter planting prior to 1991. Tables 9.1 and 9.2 show the dramatic change following the 1991 Act.

4 Crofter Forestry Experiences

The last word should be that of crofters with first-hand experience of the working of the 1991 Crofter Forestry (Scotland) Act. In their splendidly well-written book Crofter Forestry Experiences [**4**], the Scottish Crofters Union presents the views and experiences of the early pioneers of crofting forestry, who have taken advantage of the 1991 act. It also gives a "best practice" guide for the benefit of potential new crofter foresters.

By April 1996, some 235 Woodland Grant Schemes applications had been received (covering in-bye, apportioned and common land). The book described in detail the legal and practical silvicultural problems faced by and overcome by eleven crofting townships. What is so encouraging is that those involved, far from being put off by the very considerable effort required to initiate and implement the schemes, have indicated their intention to pursue other forestry schemes.

In addition to crofter forestry schemes *per se* other forestry schemes such as community woodlands and partnership schemes between local people and Forest Enterprise are described.

Examples of successful schemes are given. The potential for building up rural businesses based on forestry is emphasised. The range is large—from specialist nurseries raising plants from native seed collected locally through the education and training of people to set up contracting businesses to undertake plantation establishment, maintenance and harvesting of timber, right through to saw milling and the preparation of estate and farm timbers.

Those that against all the odds pioneered the legislation that has at last made crofter forestry possible and those of the Crofters Commission and the Forestry Authority who sold the ideas on the ground to the crofters themselves have good reason to be pleased with the outcome, as indeed do those pioneering crofters who took the lead in the early days of the green revolution in the crofting townships.

5 Acknowledgements

I wish to acknowledge the generous help given by the Crofters Commission, Inverness; The Forestry Authority, Dingwall; and the Forest Research Station of the Forestry Commission at Newton Nursery, Elgin.

The cartoon Paradise regained on the forest-croft is reproduced from the Crofting Forestry Handbook [**5**] by kind permission of the artist Chris Tyler and the Scottish Crofter's Union.

References

1. *A guide to Crofter Forestry.* The Crofters Commission, 4–6 Castle Wynd, Inverness IV2 3EQ. 1992.
2. A.L. Sharpe and S. Jacyna, *The potential for Tree Growth and Woodland Creation in the Western Isles of Scotland.* Scottish Forestry. Vol. 47. No. 4, 1993.
3. C.P. Quine and A.L. Sharpe, *Evaluation of Exposure and the Effectiveness of Shelterbelts on the Western and Northern Isles of Scotland*, Scottish Forestry, Vol. 51, No. 4, 1998.
4. *Crofter Forestry Experiences.* Aonadh Nan Croitearan (Scottish Crofters Union), Old Mill, Broadford, Isle of Skye, IV49 9AQ (address from 2002: The Steading, Balmacara Square, by Kyle of Lochalsh, IV40 8DG).
5. *Crofting Forestry Handbook.* Aonadh Nan Croitearan (Scottish Crofters Union).

Figure 9.2. Paradise regained on the forest-croft (from Ref. [**5**])

Chapter 10

Wood Production

1 The Production of Wood from Britain's Forests

One way of judging the success of Government forest policies (for example those laid down in the 1919 Forestry Act and strengthened in the immediate post-war (1939–1945) legislation), is to measure the contribution Britain's forests have made to National timber needs.

Work in this field was undertaken by the Forestry Commission in its reports to the Commonwealth (earlier Empire) Forestry Conferences of 1947, 1952, 1957, 1962 and 1968. The Forestry Commission's annual summaries of data (Forestry Facts and Figures) on production, imports and production of wood and wood products have continued this series since 1970. Apart from publishing annual statistics on forest products for every country in the world, the United Nations Food and Agriculture Organisation (FAO), with the Economic Commission for Europe (ECE) has published five full-scale, long-term European Timber Trends Studies, in 1953, 1964, 1976, 1986 and 1996, called ETTS I, II, III, IV and V, respectively, with a so-called interim review (IR) in 1969.

Early on these activities focused attention on the need to establish an internationally agreed terminology for the many different assortments of wood traded domestically and internationally and the many varied products manufactured by forest industries. In order to be able to draw up a balance sheet between the quantity of wood harvested annually and the consumption of wood and wood products, international agreement has also been sought on the quantities of roundwood required to manufacture different products e.g. cubic metres of wood required to make a ton of paper pulp.

This is a continuing programme of work due to technological changes in the industries (e.g. the increase in the use of recycled fibre in paper making) and in the nature of the wood resource (e.g. from large-sized trees from old natural forest with a high recovery of usable sawn wood to small plantation grown trees with a very much lower yield of sawn wood per unit of roundwood). When comparing the results of one study with another, difficulties can arise because historical data referring to the same period of time may alter from one study to the next. This may be due to individual countries or FAO correcting in later years errors in the original country statistics. Apart from these straightforward arithmetical errors in country data and the use of 'estimated' figures later revised to 'actual', other errors have been detected due to countries not always using the internationally agreed terminology or applying inappropriate conversion factors. Nevertheless the broad trends can be discerned with a fair degree of confidence.

In the context of production from British woodlands there is a striking contrast between the reliance on imports in the 1930s and the role British woodlands played in the seven war years, 1939–1945. Home production which met only about 4% of consumption (home production + net imports) from 1934–1938 increased dramatically during the seven war years 1939–1945. The increase was made possible only by devastating many of Britain's woodlands.

Table 10.1 Average annual home production and net imports roundwood equivalent: millions of m^3 under bark

	GB home production			Net imports[a]			
	CF	BL	Total	CF	BL	DER	Total
1934–1938	0.85	0.62	1.47	20.13	2.11	11.88	34.11
1939–1945	3.78	2.78	6.56	6.20	0.79	4.69	11.69

Source: Statement presented to the Fifth British Empire Forestry Conference 1947 by the Forestry Commission.

CF =conifers=softwoods
BL =broadleaves=hardwoods
DER = derivatives, mainly pulp and paper, veneers and plywood. The small quantities of derivative products from home sources have been included under CF or BL as appropriate.
[a]Net imports = total imports less exports: figures include direct imports to Northern Ireland and direct exports outside UK from Northern Ireland.

By 1950 the picture had changed in the sense that home production had dropped to less than 3 million m^3 and imports had increased to some 19 million m^3. At this time a number of war time controls on the use of timber were still in force, so demand was artificially restricted. Five years later when most war-time controls had been lifted, home production had increased only slightly but imports had increased from 19 to over 30 million m^3. Much of the increase was in sawn softwoods required for the post-war new housing programme.

In the 20-year period 1955 to 1975 production (removals) from British woodlands (and from 1970–1975 UK woodlands) remained at between 3.2 and 3.5 million m^3 meeting between 7.5 and 10% of consumption. Then production increased rapidly to 6.4 million by 1990 providing 13% of total consumption which by then had risen to just over 50 million m^3(see Tables 10.2 and 10.3).

As mentioned above FAO has endeavoured to publish figures on a world basis for production (removal) of wood from country's own resources and to account for the uses to which the wood was subsequently put. For the years 1939–1945, 1955, 1960 and 1965 similar studies were undertaken for Great Britain by the Forestry Commission for the Commonwealth Forestry Conferences. The studies for 1950, 1955 and 1960 are given in Table 10.4 together with the FAO data at 5 yearly intervals from 1946 to 1990. For the years 1950 and 1955 the CFC data has also been recalculated and presented using the FAO recovery factors for recovery of sawn wood from roundwood (sawn logs) rather than those used for those two years by the CFC. The FAO (R) figures include revisions of previously published figures and offer "a reasonably consistent set of data over time". Broad trends are reliably indicated by the figures. Thus there has been a big change in the importance of broadleaves (BL) as a source of industrial wood over the past half century. Annual removals of the BL sawlogs in 1990, some 0.5 million m^3, were less than a quarter of the removals in the 1940s. This in part reflects the competition from tropical timbers and from a range of temperate hardwoods from the USA and countries such as Romania and the former Yugoslavia which specialised in the production for export of consistently high quality, graded, seasoned wood. In part it reflects the legacy of the war years, a greatly diminished stock of good quality broadleaves in Britain.

There have been changes too in the use of hardwood timbers by industry since the immediate post-war years. The GB Statement on hardwoods to the 1947 Commonwealth Forestry Conference is given on page 176 and onwards.

Table 10.2 Home production and net imports roundwood equivalent: millions of m^3 under bark

	GB home production			Net imports[a]			
	CF	BL	Total	CF	BL	DER	Total
1950	0.88	2.02[b]	2.92	8.98	2.80	7.59	19.36
1955	1.14	1.85[c]	3.00	15.50	2.13	13.33	30.94

Sources: Forestry Commission Statements to 1952 and 1957 Commonwealth Forestry Conferences.

[a]See Table 10.1.
[b]Includes 0.38 million m^3 charcoal and fuelwood.
[c]Included 0.40 million m^3 charcoal and fuelwood.

Table 10.3 Home production as a proportion of total consumption of wood and wood products

Year	Publication (date)	Roundwood equivalent millions of units UK consumption			
		GB forests	UK net imports	Total UK	% from GB forests[a]
1950	CFC (1952) p.10	103 ft^3 2.92 m^3	684 ft^3 19.36 m^3	786 ft^3 22.28 m^3	13
1955	CFC (1957) p.10	107 ft^3 [b] 3.00 m^3	1093 ft^3 30.94 m^3	1200 ft^3 33.98 m^3	9
1960	CFC (1962) p.10,14	114 ft3 3.23 m^3	1286 ft3 36.41 m^3	1401 ft3 39.66 m^3	8
1965	CFC (1968) p.24,25	124 ft^3 3.5 m^3	1324 ft^3 37.48 m^3	1448 ft^3 41.50 m^3	8.5
		Forestry Facts and Figures ($\times 10^6$ m^3)			
		UK forests	UK imports	UK cons	% from UK forests
1970		3.2	40.3	42.3	7.5
1975		3.4	30.9	33.3	10
1980		3.9	34.7	36.3	11
1985		4.7	37.5	39.9	12
1990		6.4	48.5	50.2	13
1999		7.4	46.9	46.7	16

The consumption of forest products is usually calculated on a UK basis because of imported forest products landed in Northern Ireland may be re-consigned to mainland Britain and vice-versa. Consumption is apparent consumption = imports plus UK roundwood production minus exports.
The Forestry Commission's writ does not extend to Northern Ireland; the figures which it prepares for the Commonwealth refer only to mainland Britain (GB) for home production, but since 1970 its annual Forestry Facts and Figures give home production on a UK basis when calculating apparent consumption.

[a]Error in CFC Summary: correct figure is 106 to nearest million, giving 3 million m^3.
[b]% rounded to nearest 0.5 %.

GB Statement to the 1947 Commonwealth Forestry Conference

'The principal species of hardwoods considered marketable were oak, beech, common and wych elm, sweet chestnut and ash. Special trees, such as walnut and cricket-bat willow, fetched high prices and were sold separately (from other species) into specialised markets.

Other species available in such small quantities that they were of little (economic) importance, included poplar used for matches and brake blocks; birch and alder for brush backs; lime for pattern making; hornbeam and holly[1].

[1] *Author's note*: The omission of sycamore from the list of less common but regionally important species is odd, as is wild cherry (given that holly is mentioned).

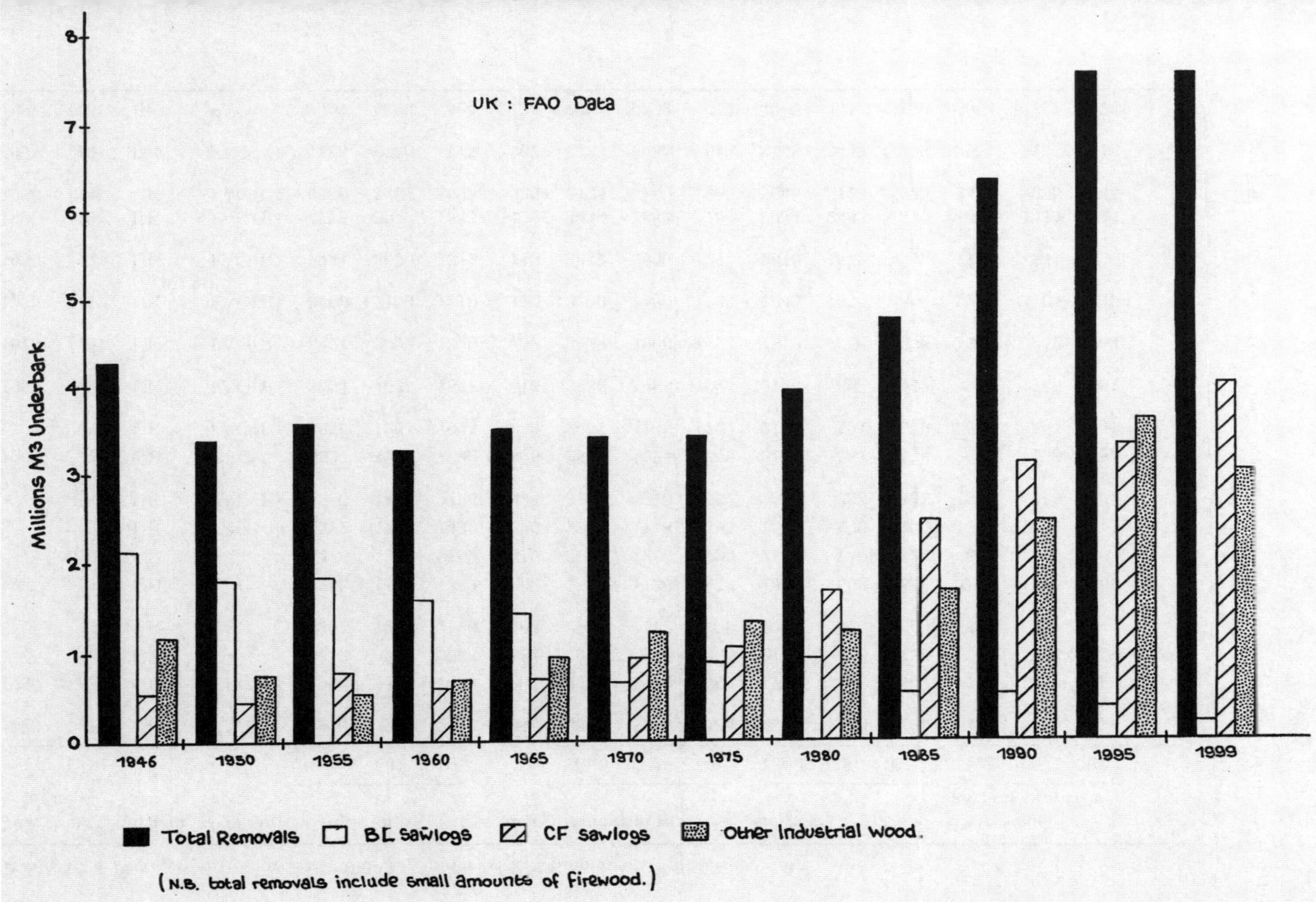

Figure 10.1. Wood Production in the UK

Table 10.4 Production (removals); ×1000 m^3 under bark. Comparison of published data

Year	Line no	UK or GB	CFC or FAO	Grand total			Fuel	Industrial wood			Sawlogs			Pulp wood and pit props[a]			(1)
				Total	CF	BL		Total	CF	BL	Total	CF	BL	Total	CF	BL	
1946	(1)	UK	FAO	4270	1776	2494	450	3820	1686	2134	2660	350	2130	1160	1156	4	
1950	(2)	GB	CFC	2908	874	1653	381	2527	873	1654	1737	335	1402	789	538	251	b
	(2a)	GB		3343	959	2003	381	2962	958	2004	2172	420	1752	789	538	251	b
	(3)	UK	FAO	3367	1247	2120	405	2962	1172	1790	2215	425	1790	747	747	–	
1955	(4)	GB	CFC	2988	1136	1852	399	2589	1136	1453	1976	590	1386	613	545	68	c
	(4a)	GB		3478	1286	2192	399	3079	1286	1793	2466	740	1726	613	545	68	c
	(5)	UK	FAO	3222	1790	1432	399	2823	1718	1105	1511	406	1105	1312	1311	1	d
	(6)	UK	FAO(R)	3566	1411	2155	434	3132	1270	1862	2614	783	1831	518	486	32	
1960	(7)	GB	CFC	3227	1237	1990	396	2830	1174	1656	2157	581	1576	673	593	80	e
	(8)	UK	FAO(R)	3256	1261	1995	399	2857	1135	1722	2170	591	1579	687	544	143	
1965	(9)	UK	FAO(R)	3502	1625	1877	416	3986	1463	1623	2140	698	1442	946	765	181	
1970	(10)	UK	FAO(R)	3421	2092	1329	392	3029	1906	1123	1813	954	659	1216	952	264	
1975	(11)	UK	FAO(R)	3400	2190	1210	110	3290	2160	1130	1945	1055	890	1345	1105	240	
1980	(12)	UK	FAO(R)	3951	2639	1312	110	3841	2609	1232	2610	1668	942	1231	941	290	
1985	(13)	UK	FAO(R)	4775	3940	835	129	4646	3882	764	2983	2463	520	1663	1419	244	
1990	(14)	UK	FAO(R)	6350	5255	1095	225	6125	5160	965	3645	3125	520	2480	2035	445	
1995	(15)	UK	FAO	7585	6780	775	262	7323	6680	643	3768	3443	325	3555	3237	318	
1999	(16)	UK	FAO	7497	6893	604	233	7264	6792	472	4210	4011	199	3054	2781	273	

Sources:
CFC Great Britain Statements to the Commonwealth Forestry Conferences 1952, 1957 and 1962
FAO World Forest Products 1946–1955
FAO(R) World Forest Products Statistics—A Ten Year Summary 1954–1963; Forest Products Statistics 1964–1974 Timber DATABASE 1964–1991
1995 Data FAO Yearbook; Forest Products 1993–1997 (published FAO Rome 1997); Total 7585 corrected to 7555 in later publications.
1999 Data ECE/FAO; Forest Products Statistics. ECE/TIM/BULL/53/2. United Nations 2001

Notes:
[a]Pulpwood and pitprops include other industrial roundwood such as transmission poles. Pulpwood includes roundwood used for wood-based panel products e.g. particle board and fibreboard.
[b]1950 lines (2) and (2a) BL pulpwood and pitprops is understood to include BL(= hardwood) sawn mining timber (page 5 GB Statement,-Note 4) Line (2a) has been adjusted to use the FAO conversion factor for recovery of sawn material from sawlogs.
[c]1955 lines (4) and (4a). Sawn mining timber included in sawlogs (GB Statement, page 5, Table 4, Note 4). Line (4a) has been adjusted to use the *FAO* conversion factor for recovery of sawn material from sawlogs.
[d]Note the major revision of line 5 FAO in line 6 FAO (R).
[e]1960 (line 7) adjusted to put sawn mining timber into sawlogs (see page 10 GB Statement, Table 7).

Oak was converted chiefly for railway wagon building, cooperage, furniture, fencing, half-timbered houses and sills, the finer butts being cut through and through to season for church work and panelling.

Wych elm was in great demand in the shipyards for boat building, while common elm was used in the same for settling blocks. Large butts of the latter were mainly converted into coffin boards and the offcuts were used for the tinplate box trade; this trade was, however, shrinking fast owing to the competitive price of fibre-board boxes. Agricultural and horticultural requirements, such as wheel hubs, wagon bottoms and barrows, were commonly made from elm.

Beech was probably the most generally used utility wood in the country. The furniture trade used large quantities, and the suitability of the wood for turnery made it popular for small tool handles, rollers and shoe heels. It was also extensively used for household wooden-ware, and for tools such as carpenters' planes and mallets.

Chestnut was used in many cases as a substitute for oak and was liked in the furniture trade because of its easy-working qualities. Ash was always much sought after, owing to its strength, toughness and adaptability to bending. It was employed for such purposes as vehicle body building, aeroplane work and handles for tools and sports goods.

Chestnut coppice was sold in the south of England when the re-growth from 15 to 20 years old for the manufacture of cleft chestnut fencing. The planting of chestnut for coppice was heavy after the 1914–1918 war, and by 1938 the price had fallen considerably. Underwood such as hazel had a limited market for pea and bean sticks, cratewood for the pottery trade, hurdles and hedging and market gardening stakes, but the supply was greater than the demand.

There was a steady demand for firewood in the country districts, and some charcoal burning was carried on.

Many old trades, such as the production of oak bark for tanning were dying out. There was still a demand for oak bark, but the price had fallen very low.'

More of these traditional markets have since disappeared or are greatly reduced. Examples include the use of wood in railway wagons; the reduction in the use of wooden coffin boards; the loss of shipbuilding as a major British industry; the extensive use of fibre-glass for boat building; the substitution of fibre-carton for all kinds of wooden boxes; the use of plastics for goods formerly made by the wood-turning industry such as brush backs, tool handles, kitchen implements; the mechanisation of agriculture with the consequent loss of wooden carts and implements with wooden handles. The furniture industry has been revolutionised by the greater use of wood-based panels, especially particle board and Medium Density Fibreboard (MDF) in place of solid wood.

Changes in coal mining methods in the post-war years created an increasing demand for hardwood sawn mining timber of the order of 0.8 million m^3 a year of sawn wood in the decade 1955 to 1965. Some 20 years later demand had dropped to half this figure, and by 1999 was negligible (Table 10.5).

Table 10.5 Use of hardwood (BL) sawn mining timber by the coal industry in Great Britain. Roundwood equivalent; ×1000 m^3 under bark

1956–1960	1961–1965	1966–1970	1971–1975	1976–1980	1986–1988
840	820	700	380	400[a]	190

Source: British Woodland Produce

Note: Sawn wood volumes converted to roundwood equivalent at 1.67 m^3 per m^3 sawnwood. Figures rounded to nearest 10 000 m^3.

[a]1981–1985 figures not given—major miner's strike.

Much hardwood sawn mining timber was cut from branchwood and stems of too poor quality to serve traditional markets.

In the 1960s the manufacture of bleached paper pulp from broadleaves was started in Britain, a decade later the plant was converted to manufacture unbleached corrugating pulp; another pulp mill was also modified to manufacture unbleached pulp from broadleaves. The report to the 1962 Commonwealth Forestry Conference reporting on annual removals of broadleaves pulpwood from British woodlands in 1960 says that "a useful outlet has been found for smaller hardwood material, derived largely from thinnings, coppice and scrub". In recent years the use of hardwood for unbleached pulp manufacture has declined but use by the chipboard (particle board) industry has increased.

In complete contrast to the decline in the production (removals) of broadleaves, production of coniferous industrial wood increased fourfold from the early 1960s to 1990. In the case of sawlogs much of the increase was due to the establishment of new sawmills geared to producing graded and seasoned constructional timbers—by far the most important UK market for sawn softwoods.

Other outlets for sawn coniferous wood include fencing (notably panel fencing). The sawn softwood mining timber market held up remarkably well with demand in 1990 little different from that in 1955 at around 0.2 million m^3 sawnwood a year. Pallets were an important outlet for the poorer qualities of sawn softwoods. Hardwoods were also used for pallets either for the blocks or for whole pallets in the case of lighter hardwoods such as poplar.

For other industrial wood below saw log size or quality, the increase is due to the establishment of pulp mills and increases in panel products production, e.g. particle board, Medium Density Fibreboard (MDF) and Oriented Strand Board (OSB). Round pitwood (pitprop) production fell during this period from nearly 0.4 million m^3 a year in the 1950s to less than a tenth of this figure by 1990.

To sum up, production of industrial wood doubled in the 40-year period 1950 to 1990 from around 3 million to some 6 million m^3 a year, during which time coniferous production rose from one-third of the total in 1950 to over 80% in 1990. There was an eightfold increase in the production of coniferous sawlogs. Production of other industrial wood increased more than threefold, most of the increase being of coniferous wood for processing into paper pulp and panel products which took

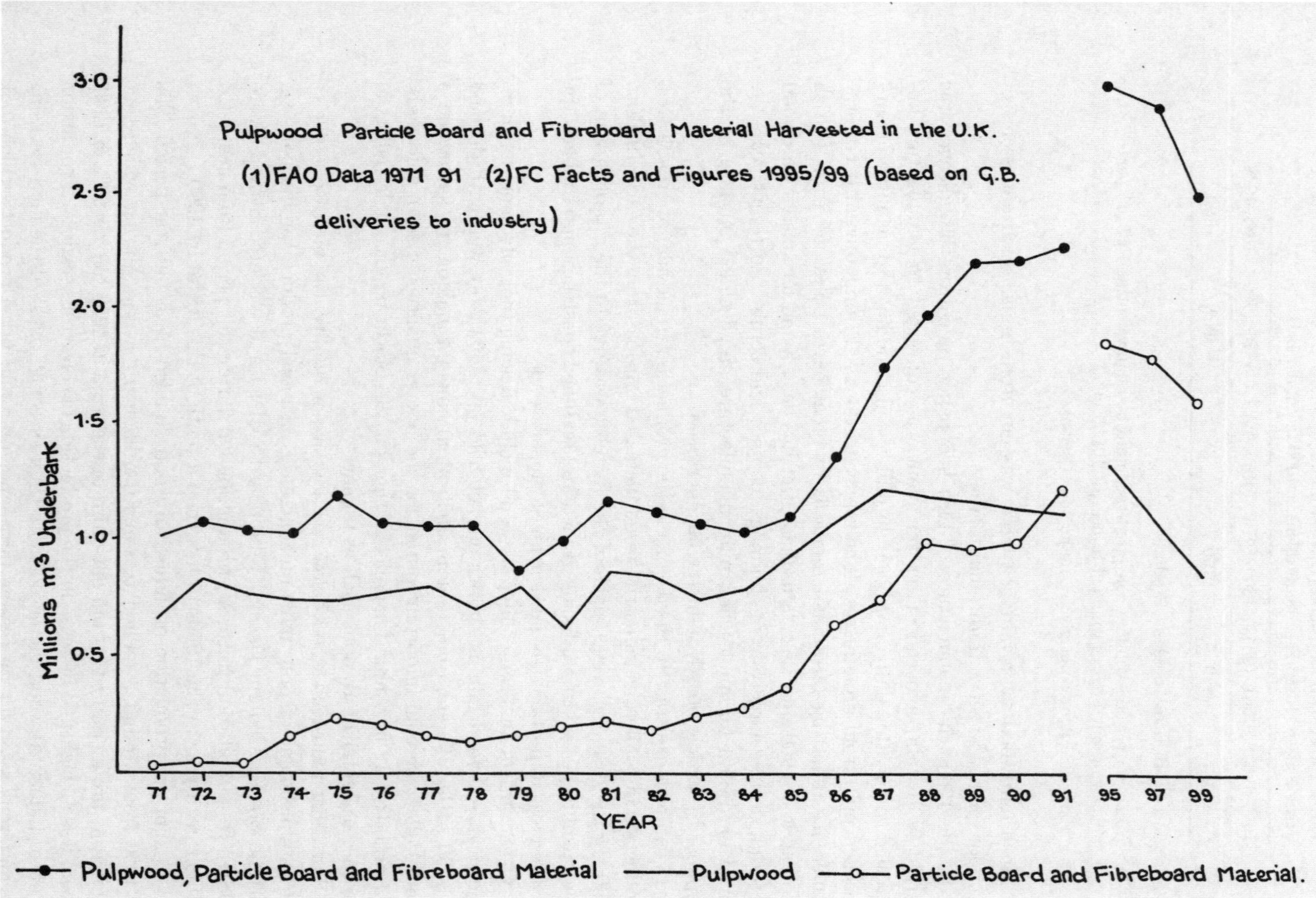

Figure 10.2. Pulpwood particle board and fibreboad material harvested in the UK

over from pitwood as the main markets for coniferous material below sawlog size (see Fig. 10.2).

By 1995 total removals had risen to 7.5 million m^3. Broadleaved industrial wood had fallen by 0.64 million m^3, whereas coniferous industrial wood had risen by 1.6 million m^3 to 6.7 million (of which CF sawlog increase was 0.31 million). By 2000 removals of coniferous industrial wood had risen by some 2.8 million m^3 to 9.5 million m^3. Broadleaves industrial wood removals fell by 0.25 million m^3.

2 Annotated Bibliography– Sources of Data

2.1 World Forest Products Statistics

1. FAO World Forest Products Statistics 1946–1955 "The most complete and authoritative world-wide statistical report on forest products for the post-war period". Published FAO Rome 1958.
2. FAO Yearbook of Forest Products Statistics 1957. Covers the years 1956 and 1957.
3. FAO Yearbook of Forest Products Statistics 1959. Covers the years 1958 and 1959.
4. FAO Yearbook of Forest Products Statistics 1961. Covers the years 1959 and 1960.
5. FAO Yearbook of Forest Products Statistics 1972. Review of the years 1961 to 1972. Published Rome 1974
6. FAO Yearbook of Forest Products 1966–1976. Covers the years 1961–1965 (average) and 1966 to 1976
7. FAO Yearbook of Forest Products 1973–1984. Published Rome 1986
8. FAO Yearbook of Forest Products 1989. Covers the years 1978–1989. Published Rome 1991

Note: To correct straightforward errors in earlier data submitted by countries; to replace preliminary "estimates" with "actual" and to comply with changes in agreed definitions and customs classifications of particular assortments of wood and wood products, figures are revised regularly so that the later the date of the publication the more accurate are the figures.

2.2 Europe, USSR, North America

1. FAO Economic Commission for Europe, Timber Committee Forest Products Statistics Published by United Nations, New York.
   Forest Products Statistics 1969–1984 published 1986
   Forest Products Statistics 1985–1990 published 1991
   Forest Products Statistics 1986–1990 published 1991
   Forest Products Statistics 1986–1991 published 1992
   Forest Products Statistics 1988–1992 published 1994
   Forest Products Statistics 1990–1994 published 1995
   Forest Products Statistics 1995–1999 published 2001
2. Forestry and the Forest Industries: past and future (Europe, USSR, North America) E.G. Richards, editor. Martinus Nijhoff Publishers for the United Nations 1987

3. European Timber Trends and Prospects FAO and the Economic Commission for Europe Published by United Nations Geneva 1953 (ETTS I).
4. European Timber Trends and Prospects: A New Appraisal 1950–1975. Published by United Nations New York 1964 (ETTS II).
5. European Timber Trends and Prospects 1950–2000. Published by United Nations, Geneva 1976 (ETTS III).
6. European Timber Trends and Prospects to the year 2000 and beyond. Published by United Nations, New York 1986 (ETTS IV).
7. European Timber Trends and Prospects: Into the 21st Century Published by the United Nations New York and Geneva 1996 (ETTS V).

UK and Great Britain

Unless otherwise stated, publications are unpriced conference documents or Forestry Commission data sheets.

1. Empire Forests and the War (1939–1945). Fifth British Empire Forestry Conference held in Great Britain in 1947. Data for Great Britain and All Empire (Commonwealth) Countries. HMSO London.
2. Empire Forests and the War—Great Britain Commentary and More Detailed Data for Great Britain London, Forestry Commission 1947.
3. Statement presented by the Forestry Commission of Great Britain to the Sixth Commonwealth Forestry Conference. Published Forestry Commission, London 1952
4. Statement by the Forestry Commission of Great Britain to the British Commonwealth Forestry Conference 1957. Published by Forestry Commission, London 1957
5. Progress Report 1955–1960 by the Forestry Commission of Great Britain to the British Commonwealth Forestry Conference 1962. Published by the Forestry Commission, London 1962.
6. Progress Report 1960–1965 by the Forestry Commission of Great Britain. Prepared for the Commonwealth Forestry Conference 1968. Published by Forestry Commission, London 1967
7. Forestry Commission—Forestry Facts and Figures: Comprehensive Coverage of Forestry Activities (State and Private) in Great Britain. Published annually since 1970 by the Forestry Commission, Edinburgh.
8. The Report of the Committee on the Marketing of Woodland Produce 1956. (The Watson Committee Report) HMSO, London 1956.
9. J.R. Aaron and E.G. Richards, British Woodland Produce. Stobart Davies 1990.
10. Russell Meiggs, Home Timber Production 1939–1945. Crosby Lockwood, London 1949.

3 Prices for Coniferous Standing Timber

In the period just before the 1939–1945 war and for some time after, the UK was Europe's biggest single importer of wood and wood products. And with home production of timber meeting an average of only about 4% of peace-time requirements between 1934 and 1938 [1], it is not surprising that the prices for British wood and wood products were largely governed by the prices paid for comparable imports.

When war-time price controls were imposed the levels that were set by Government for standing trees were considerably higher than those ruling before the War [2]. However, they remained substantially unchanged until 1947 when an increase of approximately 25% was made [3]. "A strong feeling developed on the part of woodland owners that even these prices were inadequate, on account of the increases which had taken place in wages and commodity prices generally, and prolonged negotiations with the Board of Trade took place to secure amendments. The outcome was the removal on 1st December 1949 of all price control of home-grown timber" [4].

Because of the continued reliance on imported wood and wood products—for example some 86% of total consumption in the years 1990–1994 [5] was imported—British timber prices are still largely dictated by world prices.

From the timber grower's point of view, the price for the trees sold as standing timber is of main interest to him for it is this price, less the costs of putting the timber on the market and supervising the sale, which determines the profitability of his woodlands.

Historically, most private woodland owners have sold their timber standing to timber merchants either by competitive tender or more frequently by negotiation with individual merchants. In some areas there were timber auctions of standing trees or felled logs, conducted by specialist firms on behalf of private owners or the Forestry Commission, but these were not a general feature of the home-grown market. In Kent there were regular annual auctions of chestnut coppice. Few private estates have had regular sales of timber; in part this can be accounted for by the small size of individual woodland estates and in the post-1939–1945 war period by the depletion of their woodlands during the War. Again private owners have commonly looked on their woodlands as a resource to be drawn upon as and when special needs arose (e.g. to pay death duties = estate duty = inheritance tax), rather than as annual income.

As the Forestry Commission's (mainly coniferous) woods began to yield increasing volumes of saleable thinnings in the 1950s, the Commission began to create harvesting teams to fell and convert standing trees to pitprops, pulpwood and sawlogs and other assortments for which it could find markets. Parallel with this development the Commission offered on the open market on a regular basis, increasing quantities of standing timber—mostly conifer thinnings—by competitive sealed tender.

Prices offered and accepted were kept confidential between the Commission and those tendering. Except in very unusual circumstances, the highest price was the one accepted, tenders being opened and scrutinised by small panels of office staff in the various conservancies. Whilst this system had many merits, it could result in a timber merchant losing out to a competitor for a relatively small sum of money. This might not matter too much to an individual merchant who would have bid for several parcels and expect to win enough to keep him going, it often had an unsettling effect on labour. Forest labour in the 1950s and 1960s did not have the means to travel daily to work by car nearly as far as his modern counterpart does. So, a change of timber merchant in a forest could mean that his labour force would have to work for a new firm or relocate to another forest.

The Commission therefore decided to offer longer-term contracts—up to 5 years—where the initial contract was by competitive sealed tender for the first year's cut and the prices for each subsequent year were negotiated a few weeks before the beginning of the year.

The continuity of working which resulted was widely welcomed by the Commission's field staff and by the merchants. It brought among other things a better relationship between the merchants work force and the Commission local supervisors and forest workers and with the local population, a particularly important matter in small rural communities.

It also enabled merchants to plan their marketing in a more rational and orderly manner.

With experience adjustments were made to the standard contract conditions. For example, a merchant whose rate of felling was significantly below that laid down in a contract had to pay for any increase in growth that had taken place in the trees marked for sale between the due date of felling and the actual date.

During the meetings of the Committee on Marketing of Woodland Produce (the Watson Committee) [**1**] there was considerable discussion about the merits of selling timber by auction. Professor H.M. Steven (Aberdeen University) who had wide experience of timber sales during the War as Timber Controller for Scotland argued strongly the merits of auction sales, with their transparency about prices paid (personal communication).

Those at Commission Headquarters monitoring the progress of the sales of timber in the Conservancies and the public audit office noted that with time, comparing as far as possible like with like, the prices obtained by negotiation were somewhat lower than those obtained competitively. The explanation was simple—in negotiations it was often easy for the seller to justify to himself agreement to prices offered by the buyer that were a little below the maximum the timber would fetch in a competitive sale.

After discussions with the two timber merchants associations covering respectively Scotland and England with Wales—the officially recognised home timber trade representative bodies—it was agreed that for the second and subsequent years prices (of an agreement reached initially by competitive tender for the first year) would be governed by the movement of a price index.

The index would be based on an analysis of the prices and movement over time of the prices accepted by the Commission for timber sold by competitive sealed tender. An index of price movements would be published by the Commission annually. The Secretaries of the two Associations would be given access once a year on a strictly personal and confidential basis to the relevant Commission files held at Headquarters by the Chief Marketing Officer and his staff, and only when they had satisfied themselves as to the accuracy and relevance of the Commission's calculations would the index be published.

In 1968 there was an extensive windblow of Commission and private woods—mainly in West Scotland. Special measures were taken to assist merchants to clear the windblown timber including a transport subsidy to enable merchants based in regions not affected by the windblow to harvest blown timber and transport it for processing to their mills. Sales of Forestry Commission timber outside the windblown area were reduced below their planned level.

By 1970 most of the windblown timber had been cleared and it was planned to resume the normal level of timber sales outside the area mainly affected by the blow.

Due to rises in the price of imported softwoods (conifers) in 1970 and 1971 and the desire of merchants to re-start harvesting operations in their home territory it was judged opportune to mount a large auction sale in East Scotland Conservancy. In the event the auction was a huge success, not least because it freed merchants from the uncertainties of the sealed tender system—where for example, a merchant might miss by a few £'s getting a large parcel of timber near to his processing base (e.g. sawmill/pitwood production depot).

Encouraged by Headquarters, other Conservancies followed suit and timber auctions became a regular feature of Commission and private woodland sales of standing timber and of sawlogs. An important by-product of the switch to auction sales was that it gave the Commission greater freedom to publish in detail timber prices without risk of breaking the confidentiality associated with the sealed tender or negotiated price sales. The published prices obtained by the Commission for standing coniferous timber are given below. Some implications of the prices obtained by growers—state and private—are discussed in Chapter 8.

3.1 Coniferous Standing Sales Prices in £ 1998–1999

Since 1958 the Forestry Commission has published coniferous standing sales prices and a price index based on its sales. Table 10.6 gives the average prices per m^3 over bark, adjusted for inflation, from 1958 to 1999 i.e. real average prices in £ 1998–1999.

Other things being equal, the price per m^3 is influenced not only by market conditions at the time of sale, but by the sizes of tree being sold.

Tree size influences price in two ways. First the actual cost of harvesting (felling, extracting, cutting into lengths and transporting) is substantially less per cubic metre for large trees than small trees. Secondly, the value per m^3 of larger trees yielding sawlogs is substantially more than that of smaller trees of a size yielding pitprops, panel products material, pulpwood and round fencing. Taken together tree size has thus a significant influence on standing price.

This is clearly seen in the following extract from a Forestry Commission Economics Unit Table from the 1970s (see Tables 10.6 and 10.7).

Table 10.6 Coniferous standing sales timber prices

Year ending (Financial year)	Real average price £ 1998–1999	Year ending (Financial year)	Real average price £ 1998–1999
Mar 58	28.13	Mar 81	15.00
Mar 59	28.74	Mar 82	11.42
Mar 60	27.22	Mar 83	14.03
Mar 61	27.43	Mar 84	16.97
Mar 62	25.81	Mar 85	18.51
Mar 63	21.44	Mar 86	19.30
Mar 64	21.32	Mar 87	20.98
Mar 65	21.12	Mar 88	24.24
Mar 66	21.38	Mar 89	23.36
Mar 67	21.29	Mar 90	20.87
Mar 68	19.75	Mar 91	17.58
Mar 69	21.34	Mar 92	13.50
Mar 70	20.99	Mar 93	14.37
Mar 71	22.38	Mar 94	15.75
Mar 72	21.22	Mar 95	19.34
Mar 73	23.05	Mar 96	18.05
Mar 74	43.54	Mar 97	16.78
Mar 75	45.38	Mar 98	13.70
Mar 76	26.69	Mar 99	8.44
Mar 77	28.28	Sept 99	7.65
Mar 78	25.91		
Mar 79	20.85		
Mar 80	22.19		

Source: Forestry Commission: GB Policy and Practices Division Edinburgh EH12 7AT 1999
Prices are per m^3 standing timber, overbark measure.

Table 10.7 Example of a price – size curve in the 1970s. FC data.

Conifers tree size in m^3	Price £ per m^3, standing
0.05	0.00
0.1	14.18
0.2	21.04
0.3	26.35
0.4	30.85
0.5	34.91
0.6	38.57
0.7	41.68
0.8	44.09
0.9	45.65
1.0	46.19
2.0	46.19
3.0	46.19
4.0	46.19

It follows that a quite small variation in the average size of trees sold in any one year can affect the average price obtained per unit volume and so make comparisons between one year and another not as precise a reflection of price changes per se as might be desirable for a price index.

The early sales of Forestry Commission coniferous standing timber consisted mainly of early thinnings and thus of small trees. With time, the average tree size being sold increased as the proportion of larger trees from later thinnings offered for sale increased.

Later still, larger, mature trees from clear fellings came on the market in increasing amounts. Clearly as greater volumes of larger-sized timber have been sold, comparisons of average timber prices between one year and another have been distorted. To try to overcome this problem, a technique known as Laspeyres method has been used to calculate price indices on a consistent basis. Using the Laspeyres method, "the Index reflects the price in other years of a particular size mix sold in the base year."

4 Coniferous Standing Sales Price Index (Great Britain)

Note: Figures revised in 1999 to take account of revision of the GDP deflator used by the Office for National Statistics (ONS) to the ONS GDP deflator used in the Forestry Commission Information note originally published in 1998.

Table 10.8 Comparison of the original Simple Index and the new Laspeyres Index for Great Britain (real prices) from 1971 (per m^3 overbark).

Date	Laspeyres Index at 1996 prices (Base Sep. 1996), 1971 = 100	Simple Index at 1996 prices, 1971 = 100
30–Sep–71	100.00	100.00
72	106.33	104.70
73	148.68	132.14
74	264.86	254.08
75	151.66	144.86
76	135.52	131.25
77	142.81	137.60
78	107.49	102.56
79	106.06	96.72
30–Sep–80	114.35	106.99
81	54.80	49.01
82	59.12	67.57
83	60.02	68.25
84	79.02	92.22
85	78.58	87.40
86	93.22	98.96
87	108.22	113.49
88	103.89	112.68
89	97.50	106.63
30–Sep–90	85.21	98.47
91	61.33	65.90
92	62.00	67.98
93	64.15	70.07
94	72.84	82.70
95	81.83	94.07
96	70.25	82.07
97	64.61	77.58
98	42.52	50.05
99	30.17	36.55

Source: Forestry Commission Policy and Practice Division Edinburgh EH12 7AT, 1999.

Table 10.8 and Fig 10.3 show that the Simple Index overstates price rises from 1982 (although the comparison has varied from year to year). Greater volumes of larger-sized (and thus more expensive) timber have been sold in recent years and this has moved the average price upwards. As noted above the Laspeyres Index avoids this problem because it reflects, in other years, the size mix of timber bought in the base year.

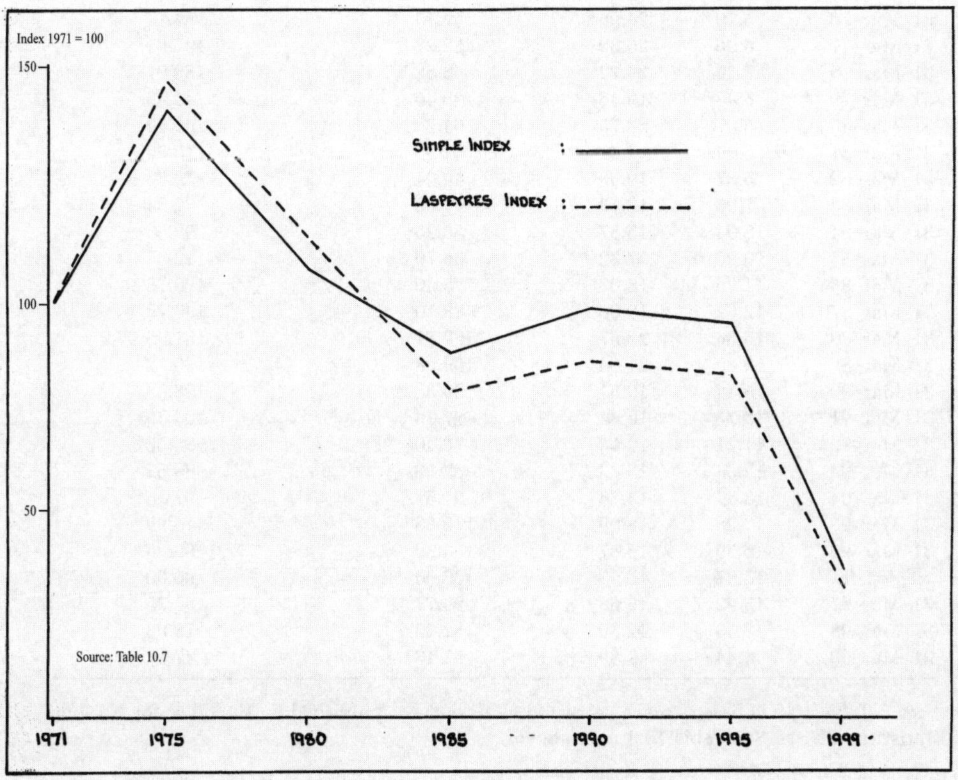

Figure 10.3. Comparison of the original Sample Index and the Laspeyres Index

Table 10.9 Coniferous standing sales prices for Great Britain (per m^3 overbark)

Date Year to	Average price £	Real average price £1996	Laspeyres Index in nominal terms September 1996=100	Laspeyres Index in real terms (1996 Prices) September 1996=100
30–Sep–71	2.49	19.46	18.22	142.35
31–Mar–72	2.65	19.85	19.38	145.04
31–Mar–73	3.11	21.57	23.76	164.61
31–Mar–74	6.31	40.73	52.75	340.68
31–Mar–75	7.86	42.44	62.32	336.27
31–Mar–76	5.79	24.96	42.75	184.17
31–Mar–77	6.96	26.39	52.22	197.91
31–Mar–78	7.25	24.22	56.32	188.07
31–Mar–79	6.48	19.46	48.40	145.40
31–Mar–80	8.07	20.75	64.22	165.12
31–Mar–81	6.44	14.01	52.95	115.20
31–Mar–82	5.37	10.68	40.01	79.50
31–Mar–83	7.06	13.13	44.63	82.94
31–Mar–84	8.94	15.87	56.25	99.86
31–Mar–85	10.26	17.32	66.70	112.51
31–Mar–86	11.26	18.03	75.40	120.78
31–Mar–87	12.62	19.60	88.70	137.79
31–Mar–88	15.36	22.67	102.71	151.58
31–Mar–89	15.80	21.84	109.08	150.72
31–Mar–90	15.14	19.51	99.38	128.08
31–Mar–91	13.75	16.44	87.33	104.39
31–Mar–92	11.21	12.63	76.40	86.03
31–Mar–93	12.33	13.43	82.23	89.62
31–Mar–94	13.87	14.73	91.51	97.15
31–Mar–95	17.28	18.09	110.85	115.99
31–Mar–96	16.59	16.87	104.60	106.36
30–Sep–96	15.97	15.97	100.00	100.00
31–Mar–97	15.92	15.68	96.77	95.35
31–Mar–98	13.35	12.80	82.33	78.93
31–Mar–99	8.44	7.89	51.13	47.81

Note: Table 10.9 is based on the corrected GDP deflator published by the office for National Statistics (ONS)—See Table 10.8 note above.

Table 10.9 shows from 1971 to 1999, for Great Britain, the average price per m^3 overbark between 1971 and 1999, the real average price in £ in 1996, the Laspeyres Index in nominal terms 1996 = 100 and the Laspeyres Index in real terms (1996 prices) September 1996 = 100.

References

1. *Report of the Committee on Marketing of Woodland Produce* (the Watson Committee). HMSO 1956.
2. Russell Meiggs, *Home Timber Production 1939–1945*. Crosby Lockwood 1949.
3. 28th CAR 1947.
4. 31th CAR 1950.
5. *Forestry Commission Facts and Figures 1994–1995.*

Chapter 11

Forestry Commission Accounts—Year Ended 31st March 1995

The Forestry Commission Accounts for the year ended 31st March 1995 cover 1994 when the Commission celebrated its 75th anniversary [**1**].

The accounts are in three parts:

- The Forestry Fund
- The Department of Forestry
- Forest Enterperise

These accounts together with an explanation of their Constitutional status is given below.

1 Forestry Fund: Financial Statements and Accounts

Foreword

1.1 *Constitutional Status*

The Forestry Commission was established in 1919. It comprises a Chairman and Board of Commissioners with statutorily prescribed duties and powers. Under legislation now consolidated mainly in the Forestry Act 1967 it is charged with the general duties of promoting the interests of forestry, the establishment and maintenance of adequate reserves of growing trees and the production and supply of timber. Since 1985, the Commission has also had a statutory duty to seek to achieve a reasonable balance, in carrying out its function, between the needs of forestry and the environment.

The Commission has the legal status of and functions as a Government Department. It reports individually and collectively to Forestry Ministers, namely the Secretary of State for Scotland (who takes the lead role), the Minister of Agriculture, Fisheries and Food and the Secretary of State for Wales, to whom it is responsible for advice on forestry policy and for the implementation of that policy.

The Forestry Fund was first constituted under Section 8 of the Forestry Act 1919 and has been continued under the Forestry Act 1967. All sums received and all expenditure incurred by the Forestry Commissioners in the exercise of their powers and the performance of their duties are included in this account. Pursuant to Section 42(4) of the Forestry Act 1967, capital sums received by Forest Enterprise from sales of land etc. are paid into the Forestry Fund and then surrendered to the Consolidated Fund, as directed by Ministers in accordance with Section 41(4A) of that Act.

The Forestry Fund accounts are audited by the Comptroller and Auditor General in accordance with Section 44(3) of the Forestry Act 1967.

1.2 *Functions and Objectives*

The Commission operated as two distinct entities with separate Departmental and forest management roles below Commissioner level:

The Department of Forestry, comprising on the one hand a Policy and Resources Group, which is responsible for support and advice to Ministers, policy development, personnel, finance, internal audit and information systems, European and international liaison, and the administration of the Forestry Commission Pension Scheme; and on the other the Forestry Authority, which is responsible for implementing the Government's forestry policy, including the control of tree felling, providing advice to private woodland owners, administering the payments of grants for approved planting, restocking and management schemes, research and the setting of standards for the forestry industry as a whole, including Forest Enterprise.

Forest Enterprise, which is a trading body responsible for the management of Forestry Commission forests. The Enterprise has a multiple-purpose remit that includes the protection and enhancement of the environment and the provision of public access and recreational facilities, as well as the production of timber.

1.3 *Accounts of Department of Forestry and Forest Enterprise*

The Department of Forestry produces an income and expenditure account detailing its net expenditure on all the services met out of the Grant-in-Aid. Forest Enterprise produces commercial accounts in a format approved by the Treasury. Both sets of accounts are audited by the Comptroller and Auditor General and published in the Forestry Commission's Annual Report.

1.4 *Review of Activities*

The net requirement for Grant-in-Aid in 1994–1995 of £76 378 000 was £12 651 000 lower than the requirement of £89 029 000 in 1993–1994. The Department of

Forestry's requirement was reduced by £793 000; Forest Enterprise's share was reduced by £10 856 000 mainly due to continuing improvements in timber prices; while the balance was to meet pensions payments which were reduced by £1 002 000 over the previous year.

1.5 *Employment Policies*

The Forestry Commission's policies in relation to the employment of disabled persons, the provision of information to and consultation with employees, and the promotion of equal opportunities, are stated elsewhere in the Annual Report.

The names of Commissioners are also shown elsewhere in the Annual Report.

FORESTRY COMMISSION: FORESTRY FUND

Receipts and Payments Account for the year ended 31.3.1995

		Previous Year
	£'000	£'000
Central Government Grants		
Grant-in-Aid	87 906	91 000
EC Receipts	12 851	5814
	100 757	96 814
Operating receipts		
Sales of timber	90 970	84 025
Other operating receipts	22 336	19 396
	214 063	200 235
Operating Payments		
Salaries and wages	62 890	63 707
Travel, subsistence and staff transfer expenses	4856	5623
Grants to private woodland owners	32 807	29 838
Other operating payments	60 805	61 636
	161 358	160 804
Surplus from operations	52 705	39 431
Other receipts/(payments) (net)	(22 845)	(16 227)
Surplus for the financial year	29 860	23 204
Appropriations		
Payments to Consolidated Fund	16 520	17 568
Excess (shortfall) of receipts over payments for financial year	13 340	5636

Statement of Balances as at 31.3.1995

Cash and bank balances		
Balance at 1 April	8036	2400
Add(deduct): Excess (shortfall) of receipts over payments for financial year	13 340	5636
Balance at 31.3.1995	21 376	8036

2 The Department of Forestry: Financial Statements and Accounts

Foreword

2.1 *Status*

The Forestry Commission is the Government Department responsible for forestry throughout Great Britain and, since 1992, has operated as two distinct entities—the Department of Forestry and Forest Enterprise.

The Department of Forestry implements the Government's forestry policy within a framework of legislation (principally the Forestry Acts 1967 and 1979, the Plant Health Act 1967, the Countryside (Scotland) Act 1967 and the Countryside Act 1968), and subject to the direction of Forestry Ministers.

The Forestry Commissioners administer the Forestry Fund which receives Grant-in-Aid drawn down from Class XIV, Vote 25. The Forestry Fund allocates Grant-in-Aid to the Department of Forestry and Forest Enterprise. These accounts cover the activities of the Department of Forestry and are prepared in accordance with a direction given by the Treasury in pursuance of Section 5 (1) of the Exchequer and Audit Departments Act 1921. They are audited by the Comptroller and Auditor General.

The transactions of the Forestry Fund and Forest Enterprise are reported separately in their respective accounts.

2.2 *Functions and Aims*

As already noted (on page 198), the Department of Forestry comprises the Policy and Resources Group and the Forestry Authority. The Policy and Resources Group is responsible for support and advice to Ministers, policy development, personnel, finance, internal audit and information systems, European and international liaison, and the administration of the Forestry Commission Pension Scheme.

The Forestry Authority is responsible for implementing the Government's forestry policy, including the control of tree felling, providing advice to private woodland owners, administering the payment of grants for approved planting, restocking and management schemes, research and liaising with local authorities and forestry and countryside bodies. The Forestry Authority sets standards for the forest industry as a whole, and monitors the performance of Forest Enterprise in the same areas of operation as those which apply to the private sector.

The aims of the Forestry Authority are to ensure, by regulation, incentive and advice, that forests and woodlands in Great Britain are protected, managed, extended and promoted in a way which best serves the national interest in accordance with Government policy.

2.3 *Research and Development*

Each year a separate report on Forest Research gives a full account of the work carried out. The quality of the research work is monitored on a continuing basis by an independently constituted Advisory Committee on Forest Research.

2.4 *Review of Activities*

Grant-in-Aid of £52 020 000 was £793 000 lower than the previous year. Net operating expenditure in 1994–1995 fell by £537 000 compared to the previous year. Whilst there was a continuing increase in the level of private woodland grant payments and in activity on advisory and regulatory functions, this was more than offset by increased income mainly from EC grants appropriated in aid and the sale of research services. Future commitments and applications awaiting approval for planting and management grants at 31.3.1995 totalled £230 million, a decrease of £21 million from the previous year's figure of £251 million.

2.5 *Employment Policies*

The Forestry Commission's policies in relation to the employment of disabled persons, the provision of information to, and consultation with employees, and the promotion of equal opportunities, are stated elsewhere in the Annual Report.

The names of Commissioners are also shown elsewhere in the Annual Report.

THE DEPARTMENT OF FORESTRY

Statement of Income and Expenditure for the year ended 31.3.1995

		Previous Year
	£'000	£'000
INCOME		
Planting grant recoveries from EC	6600	2700
Research Income	1804	1139
Sundry Income	639	470
	9043	4309
EXPENDITURE		
Grants to private woodland owners		
Woodland Grant Scheme	31 511	27 061
Other schemes (now closed to new applications)	1292	2739
	32 803	29 800
Management and Administrative Expenses		
Private woodland grant schemes	5988	5994
Licensing, regulatory and plant health services	1585	1407
Information and advisory services	6289	6277
Research and Development	10 861	10 871
Forestry Policy and Departmental Administration	6530	5510
	31 253	30 059
Net Expenditure for Year	55 013	55 550

Cash Flow Statement and Reconciliation to Grant-in-Aid for the Year Ended 31.3.1995

		Previous Year
	£'000	£'000
Net expenditure for the year per Income and Expenditure Account	55 013	55 550
Expenditure Account		
Less items not involving cash in the year		
Depreciation and loss on sale of office machinery	(1043)	(1163)
Notional employers superannuation contributions	(2700)	(2626)
Movements in net current assets	(425)	662
	(4168)	(3127)
Add Capital Expenditure	1181	397
Less Income from Sales	(6)	(7)
	(2993)	(2737)
Grant-in-Aid for the year from the Forestry Fund Account	52 020	52 813

Statement of Assets and Liabilities as at 31.3.1995

		Previous Year
	£'000	£'000
Fixed Assets		
Office machinery and equipment	3701	3469
Current Assets		
Stocks	404	368
Debtors	1153	1314
Cash in hand	1	2
	1558	1684
less Current Liabilities		
Creditors due within 12 months	833	534
Net Current Assets	725	1150
Total Net Assets	4426	4619

3 Forest Enterprise: Financial Statements and Accounts

Foreword

3.1 *Status*

The Forestry Commission is the Government Department responsible for forestry throughout Great Britain and, since 1992, has operated as two distinct entities—the Department of Forestry and Forest Enterprise.

Forest Enterprise manages the forest estate placed at the Commissioners' disposal by Forestry Ministers.

The Forestry Commissioners administer the Forestry Fund which receives Grant-in-Aid drawn down from Class XIV, Vote 25. The Forestry Fund allocates Grant-in-Aid to Forest Enterprise and the Department of Forestry. The activities of the Forestry Fund and the Department of Forestry are reported separately in their respective accounts.

These accounts cover the activities of Forest Enterprise and are prepared in accordance with a direction given by the Treasury in pursuance of section 5(1) of the Exchequer and Audit Departments Act 1921. They are audited by the Comptroller and Auditor General.

3.2 *Functions and Aims*

The functions of Forest Enterprise are to develop and manage the Commission's forests and forestry estate as a multiple-use resource, supplying timber to the wood-using industries, restocking after felling, providing opportunities for recreation, and enhancing nature conservation and the forest environment.

Forest Enterprise aims to create and maintain attractive and productive woodlands and to manage them for public benefits.

3.3 *Review of Activities*

The long-term objective of Forest Enterprise is to reach a cycle of production where its income from timber is sufficient to cover its expenditure on the forest estate, including restocking the area harvested, forest maintenance and protection, and to provide a return on funds invested. Because of the relative immaturity of the forest estate, Forest Enterprise has not reached this stage in its cycle of production.

Nevertheless, we produced an operating surplus of £10.3 million on the forest estate compared to a deficit of £0.1 million in 1993–1994. We increased net spending on recreation, conservation and heritage to £13.1 million (£12.8 million 1993–1994). We reduced the deficit on other activities to £0.9 million (£2.5 million 1993–1994). The net loss on sale of properties was £1.4 million (£1.5 million 1993–1994).

Thus, the overall deficit for the year, at £5.1 million, showed an improvement of £11.8 million from 1993–1994.

Forest Enterprise's call on Grant-in-Aid was reduced by £10.8 million to £1.2 million. The net cash inflow from operating activities increased to £13.0 million, a significant improvement from the deficit of £1.6 million for the previous year. The combination of reduced receipts from sales of properties, affected by the considerations of the Forestry Review Group, and higher capital investment increased the cash outflow on investing activities to £2.0 million compared with a cash inflow of £6.8 million in the previous year. The total net cash inflow therefore improved by £5.8 million to £11.0 million.

3.4 *Financial Objectives*

Forest Enterprise has been a set target rate of return of 3% in real terms on the forest estate and any new investment therein.

Based on individual appraisals carried out in each Forest District, the implied real rates of return during the year ended 31st March 1995 was 4% on both new planting and restocking. Other financial objectives are described in the notes to the accounts.

3.5 *Employment Policies*

The names of all Forestry Commissioners and the Forestry Commission's policies in relation to the employment of disabled persons, the provision of information to, and consultation with employees, and the promotion of equal opportunities, are stated elsewhere in the Annual Report.

FOREST ENTERPRISE

Income and Expenditure Account for the year ended 31.3.1995

		Previous Year
	£ Million	£ Million
INCOME		
Forest Estate		
Sales of timber	91.1	82.8
Other forest sales	3.0	1.5
	94.1	84.3
Recreation, conservation and heritage	3.9	3.3
Other activities	13.5	13.6
	111.5	101.2
EXPENDITURE		
Forest Estate		
Harvesting and haulage of timber	44.8	45.5
Restocking after felling	11.7	11.3
Forest roads	13.2	13.7
Forest maintenance and protection	14.1	13.9
	83.8	84.4
Recreation, conservation and heritage	17.0	16.1
Other activities	14.4	16.1
	115.2	116.6
OPERATING SURPLUS/(DEFICIT)		
Forest estate	10.3	(0.1)
Recreation, conservation and heritage	(13.1)	(12.8)
Other activities	(0.9)	(2.5)
	(3.7)	(15.4)
Surplus/(Deficit) on sale of properties	(1.4)	(1.5)
NET SURPLUS/(DEFICIT) FOR THE YEAR	(5.1)	(16.9)
Transferred to Capital Account	(5.1)	(16.9)

Balance Sheet as at 31.3.1995

		Previous Year
	£ Million	£ Million
FIXED ASSETS		
Forest Estate at Open Market Value	1394.2	1401.5
Other Tangible Assets	126.6	125.2
	1520.8	1526.7
CURRENT ASSETS		
Stocks	6.9	7.4
Debtors	18.7	17.7
Cash at banks and in hand	2.1	2.0
	27.7	27.1
CREDITORS—amounts falling due within one year	12.4	8.7
NET CURRENT ASSETS	15.3	18.4
TOTAL ASSETS LESS CURRENT LIABILITIES	1536.1	1545.1
CAPITAL AND RESERVES		
Capital Account	956.7	966.6
Revaluation Reserve	579.4	578.5
	1536.1	1545.1

Cash Flow Statement for the year ended 31.3.1995

		Previous Year
	£ Million	£ Million
OPERATING ACTIVITIES		
Receipts from sales of timber	91.0	84.0
Other receipts	19.3	17.6
Salaries, wages and related expenses	(47.5)	(49.0)
Materials, contract services and general expenses	(49.8)	(54.2)
Net Cash inflow/(outflow) from operating activities	13.0	(1.6)
INVESTING ACTIVITIES		
New Planting (excluding book value of land)	(2.0)	(2.2)
Payments to acquire tangible fixed assets	(12.7)	(8.6)
Receipts from sale of properties	12.1	17.3
Receipts from sale of other tangible fixed assets	0.6	0.3
Net cash inflow/(outflow) from investing activities	(2.0)	6.8
Net cash inflow/(outflow) before financing	11.0	5.2
FINANCING		
Grant-in-Aid	1.2	12.0
Less: Receipts from disposals programme	(12.1)	(17.3)
(Increase)/decrease in cash and cash equivalents	(0.1)	0.1
	(11.0)	(5.2)

Statement of Total Recognised Gains and Losses for the year ended 31.3.1995

		Previous Year
	£ Million	£ Million
Net surplus/(deficit) for the year	(5.1)	(16.9)
Re-evaluation surplus/(deficit) for the year	0.9	2.9
Total recognised gains/(losses)	(4.2)	(14.0)

Note: The basis of the 5-yearly valuation of the forest estate (land, timber and roads) was changed in the year ending 31 March 1993 from expectation value to open market value.

References

1. Forestry Commission. *Annual Report and Accounts 1994–1995.*

Chapter 12

British Empire and Commonwealth Forestry Conferences

1 First British Empire Forestry Conference—1920 London[1] [1]

The first British Empire Forestry Conference has been described in some detail in the Introduction. It was important to the newly formed Forestry Commission for a number of reasons.

Firstly, the holding of the Conference with its representation of countries from the whole Empire and its recognition by the Monarch himself, gave the Commission the opportunity to announce to the world at large that the Government of the UK

[1]The Conference titles and locations (e.g. London, Great Britain, UK) are taken from the respective official reports.

had at last formulated and promulgated a forestry policy and through the Forestry Act of 1919 and the setting up of the Forestry Commission had established the means to implement that policy.

Second, it gave the Commission an immediate international role and presence in an Empire that encompassed a large part of the world's forest resources—both temperate and tropical.

Third, and by no means least, it enabled the Commission to draw upon the expertise of a number of forest-rich countries in formulating the formidable programme of work necessary to create a substantial reserve of timber in the UK, a country that had neglected its forests over many decades.

Early Conferences—which lasted for several weeks (!)—followed a standard procedure.

- The Conference was convened by a Standing Committee of Empire Forestry Conferences (established at the second conference in Canada in 1923), which continued to function between conferences and which decided on the topics to be discussed, authors of key papers and laid down the content of and format of Country Reports which updated the key forestry and forest products trade statistics for each member country.
- Plenary Sessions discussed and approved the Resolutions drafted and presented by the technical committees.
- Technical Committees dealt with the main subjects under discussion such as forest policy; education; forest inventories; production of and international and Empire trade in timber and other forest products; forestry research; forest products research; and the provision of central services to member countries (e.g. forestry abstracts).
- Extensive field trips were organised, where the host country was subjected to a kind of high powered technical audit of its forests and institutions by the members of the conference.

The Conference resolutions were conveyed informally by delegates direct to their Governments and transmitted formally by the Standing Committee to Governments. Similar procedures are still followed by the shorter more stream-lined conferences of recent times.

2 Second British Empire Forestry Conference—1923 Canada [2]

The second British Empire Forestry Conference in Canada lasted from July 25th to September 7th—a little over 6 weeks. During this time, the Conference met twice in full plenary session and in committee on 15 occasions during the extensive tour that covered all of Canada, except for Prince Edward Island and Nova Scotia.

The Proceedings, Resolutions and Brief Description of Tours run to 416 pages. The Conference Resolutions covered much the same ground as the first Conference:

2.1 Forest Policy

Each country should lay down a forest policy aimed at producing a sustained (=sustainable) yield of timber and other forest products (very important in tropical rain forests) whilst maintaining and improving climate conditions in the interests of agriculture and water supply.

2.2 Softwood Resources

There was a need to conserve and to augment coniferous timber resources in view of the great and increasing drain on world softwood forests.

2.3 Empire Trade in Forest Products

Active steps were needed to maximise international timber trade between Empire countries (part of the general policy of 'Empire preference' for all goods).

2.4 Education

Need for a central Institution at Oxford University to provide post-graduate training in forestry and research for the Empire (repeat of the 1920 resolution).

2.5 British Empire Forestry Association

The incorporation by Royal Charter of this Association made up of and supported solely by private individuals was warmly welcomed. Its journal was regarded as a major link between foresters in all countries within the Empire.

2.6 Empire Forestry Bureau

It was recommended that a Bureau should be established and paid for by Governments. It would be the official means of disseminating technical data and information generally from world-wide sources to Commonwealth countries.

2.7 A Standing Forestry Committee

Needed to ensure continuity of action by and between sessions of the British Empire Forestry Conference.

2.8 Investigation Into Forest Products

Need to strengthen and extend the work of the (British) Forest Products Research Board and to establish in Great Britain a well-equipped Forest Products Research Laboratory.

2.9 Forest Fire Protection, Canada

There was a need to strengthen fire protection in Canada.

2.10 Silviculture in Canada

Canada should strengthen silvicultural methods in Canadian forests (directed at Government and logging companies).

2.11 Shifting Cultivation

Problems of effect of shifting cultivation on forests in several countries must be addressed.

3 Third British Empire Forestry Conference—1928 Australia and New Zealand [3] Fourth British Empire Forestry Conference—1935 Union of South Africa [4]

The third and fourth Conferences held in Australia and New Zealand (1928) and the Union of South Africa (1935) both re-affirmed Resolution No. 1 of the First Conference (1920), re-affirmed at the Second Conference. Stripped of its conference trappings, this key resolution called for countries:

- To formulate a forestry policy based on an inventory of its forest resources and possibilities for expanding its forests.
- To establish the Government mechanisms for financing and implementing that policy.

On education, the third Conference reported on the successful start made by the Imperial Forestry Institute and on the desirability of putting it on a permanent basis.

On the silvicultural side more research was called for on the influence of forests on climate, water conservation and erosion; and the inter-action of tropical agriculture and forestry.

The increase in inter-Empire trade in timber was welcomed, as was the increase in forest products research in member countries. For small countries that could not justify a forest products research laboratory it was recommended that Britain's Forest Products Research Laboratory be allowed to take on—on repayment—work for them.

4 Fifth British Empire Forestry Conference—1947 Great Britain [5, 6]

The Fifth Conference held in Great Britain from 16th June to 19th July 1947 was an important one in that it carried out an assessment of the contribution Empire forests made to the 1939–1945 war, and re-asserted the importance of Empire Forestry Conferences to the countries of the Empire. Some 109 senior foresters and timber trade representatives attended as delegates, plus one honorary delegate representing the Royal Burgh of Inverness and three guests—the Director of the American Forest Products Laboratory in Madison and two representatives of the Forestry Division of the Food and Agricultural Organisation (FAO) of United Nations.

The Conference was opened by the Minister of Agriculture and Fisheries and the Secretary of State for Scotland, supported by the Secretary of State for India, the Secretary of State for the Colonies, two Forestry Commissioners and the Chairman of the British Empire Forestry Association. It was chaired by the chairman of the Forestry Commission, Lord Robinson, and messages were exchanged with His Majesty the King. In addition to the UK, countries represented were:

Australia	India	North Borneo
Basutoland	Ireland—Eire	Palestine
Bechuanaland	Jamaica	Sarawak
British Guiana	Kenya	South Africa
British Honduras	Malayan Union	Southern Rhodesia
Burma	Newfoundland	Swaziland
Canada	New Zealand	Tanganyika
Cyprus	Northern Rhodesia	Trinidad
Fiji	Nigeria	Uganda
Gold Coast	Nyasaland	

Thirty-six Forest Authorities presented statements on Empire Forests and the War (printed by HMSO as a separate publication from the Conference report). Noting that in all countries forest authorities had to turn from growing trees to harvesting timber, the publication gives—by countries—the land under forest, volume of standing timber, increment, output of home-grown wood and of imports, exports for pre-war and each war-time year and the effect of war-time exploitation on the forest resource.

The need for timber in the pursuance of war is well emphasised by the fact that the total strength of Military Forestry Units producing wood in Britain reached nearly 10 000 men, of whom 6800 were Canadians, plus civilian units from Newfoundland and British Honduras.

In nearly every theatre of war except North Africa (where there are no forests) special forestry companies of Royal Engineers, or their equivalent in other allied forces, helped to supply the allied armies with timber.

The Conference passed 14 Resolutions of which the key points of 11 resolutions are noted briefly below (the other 3 resolutions were concerned with the administration of future conferences):

1. The Forest Policy resolutions of the first (1920) Conference were reiterated and countries which had not taken substantive action to adopt sound policies along the lines of the 1920 resolution were urged to do so.
2. Land use planning by countries was advocated so as to make best use of the capacity of soil and climate for its most appropriate function, including forestry. Governments should ensure that private forests are properly managed (tax regimes may need to be adjusted to help owners in this respect).
3. Countries should assess their forest resources using aerial survey where applicable.
4. All forest areas should be managed under working plans approved by the highest competent authority.
5. Forest products research should be strengthened; a Forest Products Committee should be set up to ensure higher, more uniform standards and closer international co-operation.
6. There should be determined efforts to promote better utilisation of timber and the avoidance of waste (e.g. by the allocation of the poorer qualities and smaller sizes of timber to the pulp and board industries which could make good use of them).
   To this end there should be closer contacts between producers and consumers to revise specifications (by implication downwards?) which could lead to better utilisation of the timber resources available.
7. At all levels education should be strengthened; courses should provide on a scholarship basis to enable sub-professional persons of outstanding ability to enter professional grades.
8. The Imperial Forestry Institute should be better supported to enable it to carry out the educational and research work for which it was founded.
9. The Conference noted the establishment of the Imperial Forestry Bureau (whose information service operated "most efficiently" through the regular publication of Forestry Abstracts, including Forest Products and Utilisation, and a News Bulletin).
10. The Empire Forestry Association was warmly commended for promoting greater interest in forestry throughout the Empire and for undertaking the preparation of an Empire Forestry Terminology.
11. Co-operation with FAO was warmly welcomed as an on-going activity.

5 Sixth British Commonwealth Forestry Conference—1952 Canada [7]

The Sixth Commonwealth Forestry Conference was held in Canada. It was the first to be held under the new title Commonwealth Forestry Conference, but the numbering of conferences continued to follow the original sequence. Twenty-two Commonwealth countries were represented.

The UK delegation was led by the chairman of the Forestry Commission, Lord Robinson, who sadly died in Canada. The other UK delegates included the Director General of the Forestry Commission, the Director of the Forest Products Research Laboratory, the Forestry Adviser to the Colonial Office, members of the Timber Trade, the Universities, the Commonwealth Forestry Bureau and the Empire Forestry Association.

The main subjects discussed were those of the previous (1947) conference, with the main emphasis on forest management, conservation of timber resources and the utilisation of wood.

6 Seventh British Commonwealth Forestry Conference—1957 Australia and New Zealand [8]

The seventh Conference was held in Australia and New Zealand from 26th August to 10th October 1957, with a similar level of UK representation to that of the sixth Conference.

Large-scale conifer afforestation carried out in South Australia and in New Zealand was of particular interest to the delegates from Great Britain and was one of the principle items of the Conference. The other subjects studied and debated were those of previous Conferences.

7 Eighth British Commonwealth Forestry Conference—1962 East Africa [9]

The eighth Conference took place in East Africa (Kenya, Tanganyika and Uganda) between 25th June and 28th July 1962 and was attended by more than 60 delegates from 21 countries of the Commonwealth.

It was different from previous conferences in that it took place at a time when the African countries of the British Commonwealth had gained, or were in the process of gaining their independence. Some of the main conference resolutions were therefore addressed to newly independent countries. It is reasonable to suppose that far from these resolutions being imposed on these new countries by the older countries, they were actively sought by foresters and others of the new countries concerned, who wished that the forestry and land use policies in existence before independence, should continue after independence.

Although the Conference discussed the future role of plantation forestry, there was also an emphasis on matters not always directly associated in people's minds with "commercial" forestry such as:

- the reconciliation of wildlife conservation with forestry
- forests and water conservation

and with the strong direct dependence of large sections of the population on forests for their very existence.

With these three subjects very much in mind, the President of the Conference, the Minister of Lands, Forests and Wildlife in the (now independent) Government of Tanganyika included the following in his opening speech:

"All previous Conferences have been held in what I might call the more advanced Commonwealth countries. Now, for the first time, the Conference assembles in East Africa, where one country, Tanganyika, has recently become independent, and the others are on the threshold of independence. These countries are still in the early stages of development. For the most part cultivation is by hand, and industrial development cannot be compared with that of advanced Commonwealth countries. In these circumstances, the peoples of the East African countries are in many ways more intimately connected with forests and forestry than the peoples of those countries where this Conference has previously met. Cooking is largely done with firewood; dwellings are mainly constructed of poles and the land for cultivation is laboriously cleared with the axe and panga from the forests and woodlands which still cover extensive tracts of country. Large sections of the population depend upon the natural forests for such everyday requirements as tool handles, rope and string, wrapping materials, weaving fibres, thatch, medicine and many items of food including the meat which they get from hunting, and honey from wild bees. While the Conference is discussing scientific forestry of a very advanced kind, I hope it will not forget the many millions of people in the Commonwealth who live in the way I have described.

It is appropriate to the East African setting that the Conference should have among the first items of its agenda the reconciliation of wildlife conservation with forestry and the question of forests and water conservation. These are two matters which are very real practical issues in our forestry here in East Africa, and delegates will have an opportunity of seeing examples of them during their tours." The President concluded by recognising the important role of plantation forestry in the region.

"So much of our East African forestry depends upon the introduction of useful fast-growing exotic trees that it is encouraging to see that this subject also has an important place on the agenda. Along with it goes discussion on the comparatively new science of tree-breeding which seeks to provide high-quality stock for tree plantations.

These items have an important bearing on the struggle to remove poverty, ignorance and disease throughout East Africa, since adequate supplies of good, cheap wood are essential for the improved standards of living which we so greatly desire for our peoples."

The 17 pages of resolutions dealt first with the forestry situation in Kenya, Tanganyika and Uganda that would arise in the period following their independence. The subjects commented on included forest policy; the expansion of forests by planting; forest management needs including wildlife management; the need for a regular review of and forecast of wood production and consumption; staffing; education; and research.

Resolutions affecting all Commonwealth countries dealt with forest policy and management (including the need to safeguard tropical high forest); the role of private, communal and state forestry; education and training of staff; education of the public in forestry; technical silvicultural practices including those for fast growing exotic plantations and those for retaining and improving savanna woodlands; utilisation; forest protection; the roles of the Commonwealth Forestry Institute (Oxford) and the Commonwealth Forestry Bureau.

8 Ninth Commonwealth Forestry Conference—1968 India[2] [10]

The 9th Conference was held in India from 3rd to 27th January 1968. The Vice-President of India in opening the Conference said that the Indian setting to the Conference had its own significance. The very foundations of the ancient Indian culture were laid in hermitages under the shadows and influences of the surrounding forests. He went on:

> "It is this kind of influence that surely must have impelled Gautama Budda to say his famous words:
>
> - The forest is a peculiar organism of unlimited kindness and benevolence that makes no demands for its sustenance and extends generously the products of its life activity; it affords protection to all beings, offering shade even to the axe man who destroys it."

The Conference was important in that it discussed at length the FAO forecasts—presented in the opening key note paper by FAO—of a substantial increase in world consumption of wood and wood products to 1975; this trend was expected to continue through succeeding decades and to create increasing difficulties of meeting demand. The challenge for governments was to meet the needs of people for wood and wood products through "production forestry" and the "social benefits" the forest could confer on peoples—protection, recreational and other non-wood benefits such as medicines.

Meeting the enormous demand for fuelwood by the large rural population of India was a matter of concern to the Indian government.

The use of cattle dung—for many the only alternative source of energy to wood—also gave rise to concern as the dung should be used as fertiliser to increase the yield of locally grown food.

[2] British dropped from the title at this and all subsequent Commonwealth Forestry Conferences.

The 17 pages of resolutions covered much the same ground as the previous Conference but with greater emphasis on the role of plantation forestry as a means of increasing wood supplies through the use of fast-growing species—mainly conifers; the need for greater research into the utilisation of plantation grown exotics (especially small-sized wood from thinnings) and the relevance of the rapidly developing particle board (wood chipboard) and fibreboard markets to their utilisation.

The need to protect the tropical and other natural forests was also highlighted and an old Indian saying was quoted by the Vice-President as a warning if social consciousness did not impose some restraint on peoples:

> "Forests precede mankind and deserts follow."

9 Tenth Commonwealth Forestry Conference—1974 United Kingdom [11]

The 10th Conference held in the UK from 4th–27th September 1974 was attended by some 226 delegates. The largest of the delegations of the 21 countries represented at the Conference was from the UK (Great Britain and Northern Ireland)

UK forest services	37
Other official bodies and universities	48
Companies	12
Private individuals with forestry interests	7

Australia	13	Malaysia	5
Brunei	2	Mauritius	1
Canada	26	New Hebrides	1
Fiji	2	New Zealand	5
Ghana	2	Nigeria	10
Hong Kong	1	Papua New Guinea	3
India	4	British Solomon Islands	2
Jamaica	1	Tanzania	1
Kenya	4	Trinidad & Tobago	2
Lesotho	1	Uganda	4

In India the first Plenary Session had broken a Conference precedent by being addressed not by a traditional forester dealing with forestry policies, but by a leading forest economist dealing with forecasts of world demand for wood and supply possibilities. In the 1974 Conference, the first Plenary Session was devoted to the Conference theme, the Forest and Global Environment, and the paper was read not by a forester, but by Sir Frank Fraser Darling, one of the world's outstanding ecologists [**12**].

The five technical committees of the Conference dealt with the formulation of national forest policies; forest management (a) the influence of management practices on the environment and (b) improving the efficiency of forest management;

harvesting and market planning; and the optimum utilisation of forest products.

Although not expressed so directly as reported here, Conference recommendations (note change from Conference resolutions used in all previous Conferences) stressed the need to analyse the long-term effects of forest policies and working methods on the environment and to take these into account as well as short-term cost effectiveness in deciding which policies and working methods to adopt.

10 Eleventh Commonwealth Forestry Conference—1980 Trinidad and Tobago [13]

The 11th Conference was held in Trinidad and Tobago from 8th–25th September 1980, with a post Conference tour in Jamaica from 26th–29th September; the theme was "Forestry's Contribution to Social and Economic Development." The Republic of Trinidad and Tobago lies close to the mouth of the Orinoco river (Venezuela) and was at one time joined to the mainland of South America. The tropical forests and much of the wildlife (e.g. the butterflies) of mainland South America are found in Trinidad and Tobago. Since the early part of the century teak and Caribbean pine have been planted to increase wood production.

Imports of wood and food have been financed by the export of oil from Trinidad's oil fields but the need to make more use of the island's agricultural and forestry potential has been recognised for many years.

Conference recommendations included the need to increase the pace of reforestation; the desirability of promoting agro-forestry in situations where the production of fuelwood and food is beneficial from both a social and economic point of view. The Conference stressed that foresters and forestry organisations should continue to play a leading role in conservation of the world's natural resources. It noted that greater efficiency in the management, harvesting and utilisation of trees and of the secondary non-wood products of the forest, can result from better international co-operation, including the support of and use of centres of specialised excellence.

11 Twelfth Commonwealth Forestry Conference—1985 Victoria, Canada [14]

The 12th Commonwealth Conference was held in 1985 in Victoria, British Columbia, Canada from September 8th–22nd. The theme was "Investment in Forestry—the needs and opportunities." The Conference discussed what needs to be done to secure the many and diverse social, economic and environmental benefits of forestry by well-planned investment.

There were 180 registered participants from 28 Commonwealth countries, the United States of America and 10 international (including Commonwealth) organisations. The Conference report notes that there was a significant departure from previous conferences in that the Conference duration was reduced from three to two weeks and was arranged around a single theme.

The 373 page report of the proceedings included a "Concluding Statement and Recommendations" which reads:

"There has never been a time in which the international community has been more conscious of the importance of forests. The Food and Agricultural Organisation of the United Nations (FAO) has declared 1985 to be the International Year of the Forest. It did so because of its awareness of the need for better management of forests and of the dangers posed by the rapid depletion of tropical forest resources. The Ninth World Forestry Congress, which met in Mexico in July, focused on two major threats—forest degradation in the tropics and the threat of air pollution to some temperate forests.

Discussions at the Twelfth Commonwealth Forestry Conference in British Columbia, Canada, in September, 1985 also concluded that the depletion of forest resources in many parts of the world is a threat to the well-being and often the survival of large parts of the world's population.

The depletion of forests takes different forms. In the tropics much forest land is cleared for agriculture as a result of increasing population and the relentless search for fuel and forage. In some countries, temperate and tropical, forest depletion is a result of industrial exploitation.

The scale of the problem is large. FAO estimates that tropical forest is being cleared at some 11 million ha each year and as much as 40% may have been lost in the last 25 years. In Canada in 1983 it was estimated that coniferous forest was being lost to non-renewal, fire, pests and disease at about 1 million ha per year. The challenge for the future is underlined by the estimate that between the year 1980 and 2025, the total number of people on earth will double from 4.4 billion to 8.5 billion.

Both in the countries faced with the critical needs of rising populations, and in countries faced with rising public awareness and expectations concerning the quality of the environment, it becomes essential that forests are managed more wisely and in a more integrated fashion than hitherto, under policies that recognise the multiple benefits that forests can provide.

A vital question before us is how to conserve and, where possible, renew the forest resource in a sustainable manner. Forest renewal has been a dominant theme throughout the Conference; for too long we have capitalised on, and drawn income from the forest with inadequate reinvestment. The concept of responsible stewardship of the forest resource is a key one.

Such stewardship relates particularly to forest conservation and forest renewal but it must also relate to expanding the resource by new plantations and tree planting outside forest lands. All such new planting needs to be closely linked with improved and sustainable agriculture and must incorporate community forestry and agroforestry.

Forestry is no longer an activity concerned with specially designated forest land remote from the public. Society and other land users are now intimately involved and increasingly the forest manager is concerned with achieving a wise overall use of land which balances the needs of wood and food production and environmental quality.

The challenge is one of setting national priorities and objectives and of optimising the use of resources both physical and human. To achieve the required action on the ground will require a major investment over the next decades, particularly in developing countries. This will require great versatility and professionalism by foresters, at least in giving the highest quality advice to politicians and to international agencies and to business interests to ensure that the necessary investment decisions are made.

With this background, the Twelfth Commonwealth Forestry Conference recommends to Commonwealth governments:

11.1 Public awareness

That vigorous steps be taken immediately to raise the level of public awareness and comprehension of the broad spectrum of forestry to encourage well-informed and rational debate on forest management issues. Forest managers should ensure integration of timber production, environmental responsibilities and the varied needs of society. They should become involved in the process of land use planning and take full account of the social needs of the people they serve.

11.2 The environment

That public awareness and the political significance of environmental issues, notably landscape, recreation and nature conservation, are now major factors to be considered. If forestry policies are to reflect the changing needs and aspirations of society they must be based on multiple-use principles with the appropriate balance of environmental and financial objectives. To achieve this requires employment of a wide range of disciplines and investment in improved communications and skills both in environmental and financial management.

11.3 Forest regeneration and management

That there is an urgent need for new initiatives to control the rapid rate of forest depletion in many countries and to improve the standard of regeneration and protection of forests after harvest. The most urgent requirements are to:

- Reform stumpage and wood pricing policies with the aim of increasing financial resources for forest renewal;
- Require investors and users to meet specified forest management objectives in exchange for security of tenure.

11.4 Pressure on tropical forests

That pressures on natural forests in tropical and subtropical countries be relieved by promoting integrated rural development plans including:

- Establishment of new plantations to increase supplies of timber for industrial use, fuelwood and other purposes;
- More efficient and sustainable systems of agriculture including agroforestry.

11.5 Land degradation

That in view of the rapidly deteriorating condition of forests, soil and water resources and the widespread land degradation and desertification in many developing countries, that current action programs be expanded and that the socio-economic and ecological consequences of failing to take urgent action be evaluated and publicised. The guidelines for increased investment should follow the FAO five-point action program for tropical forests and the Report of the World Resources Institute, which call for a doubling of international assistance in the next five years.

11.6 Rural development and fuelwood

That forestry agencies actively promote the role of forestry as an integral part of rural development and convince their governments to accord high priority to forestry projects including small enterprises and in particular to address the fuelwood crisis in developing countries.

11.7 Financial management and economic analysis

That forest enterprises adopt and further develop appropriate management tools of a financial and economic nature, including methods of investment appraisal and accounting, that present a true and realistic picture of forestry investments. In order to secure funds, project proposals must include costs and benefits which are clearly quantified in both economic and financial terms and subsequent performance in achieving objectives must be audited and reported regularly. A working group should be formed to develop the subject.

11.8 Training needs

That governments, international aid agencies and forestry training institutions review their existing training programs and invest as necessary to ensure that an adequate number of professionals, technicians and forest workers are trained and that there are satisfactory career paths developed. Special attention should be given to adapting the scope of training to the changing needs of forestry.

11.9 Research

That to ensure the most effective use of available funds, priorities for research should take account of practical relevance and prospects for application. Particular attention should be paid to those activities where expanded investment is required, notably forest renewal and forestry in relation to agriculture and rural development, especially for dry lands. In addition, special attention should be paid to integrated methods of analysis of management systems designed to ensure that any series of operations is, in aggregate, optimised in financial and business terms.

11.10 Use of information

That, in view of the widespread failure of land use managers to make use of available scientific and technical information, they should review the extent of this failure and invest as necessary in improved information systems.

11.11 High technology land-use planning

That forest administrations examine the use of satellite imagery and other remote sensing information and subject to cost-effectiveness, consider using this in planning, mapping and monitoring.

11.12 Commonwealth co-operation

That ministers responsible for forestry in Commonwealth countries should take appropriate action to support and strengthen co-operation in forestry, particularly relating to research, training and information exchange.

11.13 Commonwealth institutions

That the future of Commonwealth Forestry Conferences and of the activities of Commonwealth institutions be reviewed and strengthened in the light of the changing needs of member countries."

12 Thirteenth Commonwealth Forestry Conference—1989 Rotorua, New Zealand [15]

The 13th Commonwealth Forestry Conference was held in New Zealand from 17th–30th September 1989; its theme was "Forestry—a Multiple-use Enterprise." More than 200 delegates from 33 countries attended.

The Conference press release drew attention to the fact that the Tropical Forestry Action Plan initiated by foresters in 1985 had drawn attention to the fundamental social, technical and institutional causes of tropical deforestation viz:

- clearing for agriculture—the main cause—to meet the basic food requirements of rapidly increasing populations;
- the massive demand for fuelwood and forest grazing and forest fires.

The Conference stated categorically:

- contrary to popular belief, commercial logging is not the major cause of deforestation.

The Conference drew the attention of Commonwealth Heads of Government (due to meet in Kuala Lumpur in October 1989) to the effect of rising carbon dioxide concentrations on global climate (global warming with all its potential dangers) and to:

- the contribution forests and wood-based products make to the storage of carbon;
- the fact that the manufacture of forest products requires less fossil fuel and hence releases less carbon dioxide than the manufacture of many other substitute materials.

The Conference therefore recommended that Heads of Government take steps to institute programmes and incentives to promote both forest restoration and afforestation and to encourage a greater use of wood products, and to take the steps necessary to promote and then act on a special Commonwealth Tropical Forestry Action Plan to make a positive contribution to the world-wide Plan as a whole.

Other recommendations included those dealing with training and research; the formulation and promotion of policies and where necessary laws to encompass all forest values and to ensure active participation by the public and especially local communities in discussion and implementation of these policies

13 Fourteenth Commonwealth Forestry Conference—1993 Kuala Lumpur, Malaysia [16]

The 14th Conference, with the theme "People, the Environment and Forestry—Conflict or Harmony", was held in the context of the adoption by all Commonwealth countries of the recommendations of the United Nations Conference on Environment and Development (UNCED Rio de Janeiro 1992).

13.1 Statement of Forest Principles

The keynote address was "Sustainability in the Forest".

The Conference made six recommendations to governments, four to forest services and two to research organisations. Six recommendations to the Standing Committee on Commonwealth Forestry dealt with administrative matters with which we are not here concerned.

A final recommendation was addressed to Commonwealth Ministers responsible for forestry.

The Conference recommendations are given below.

13.2 Recommendations

This Conference,

- Recognising the commitments made by Commonwealth countries at UNCED;
- Recognising the concern about the sustainable development of forests in Commonwealth countries;
- Recognising that sustainable forest development and the need for sustainability criteria are matters applicable to all types of forest;
- Recognising the need to balance the economic use of forest resources and the protection of the environment to benefit present and future generations; and
- Recognising the need for continued human resource development in the fields of forest education and training,

Recommends:

1. TO GOVERNMENTS
   (a) That Commonwealth countries review their forest legislation, policies and strategies that have an impact on forests on a regular basis to ensure that they meet current and future needs at local, national and international level.
   (b) That governments recognise the potential of forests to provide a multiplicity of goods and services, the rapid increase in demand for such goods as pharmaceutical compounds and services such as nature-based tourism, and the need to establish mechanisms that ensure that the use of forests for these purposes generates revenue and other benefits.
   (c) That governments and financial institutions should be made aware that appropriate investment in new forest areas and in the management of existing forests can contribute significantly to the amelioration of the world's environmental problems.
   (d) That governments devote more attention to the relationship between forestry and climate change, particularly:
      i. by increasing the levels of research and public awareness of the interaction between forests and climate;
      ii. by encouraging the use of forests, both natural and man-made, for carbon sequestration;
      iii. by exploring the potential to develop energy systems based on renewable forest resources;
      iv. by encouraging the more efficient use of wood-based products in preference to products requiring high-energy inputs.

(e) That there is a need for more effective community participation in forestry decision-making processes.

(f) That in the development of forestry policies, management and initiatives, gender-specific issues be addressed where relevant; and that organisations identify and remove barriers to the participation of women in forestry initiatives and the profession of forestry.

2. TO FOREST SERVICES

(a) That forest managers should recognise the need to take account of the political dimension and to be pro-active in providing sound information and advice to the community and decision makers.

(b) That, in the implementation of forest management systems, specific attention be given to the conservation and enhancement of biodiversity.

(c) That, in order to ensure political and budgetary support, sustainable forest development practices much be in keeping with overall national development objectives and international obligations. This may require the formulation of appropriate guidelines or codes of practice for forest management operations.

(d) That exemplary forest sites should be used to demonstrate the value and principles of sustainable management, and encourage country initiatives in this area.

3. TO RESEARCH ORGANISATIONS

(a) That forest research and development be intensified, particularly for better understanding of forest biodiversity, and that new developments in fields such as biotechnology, geographical information technology, community participation, and valuation of non-market benefits be encouraged.

(b) That, for maximum benefit to countries with common problems, research should continue to be the subject of international collaboration, exchange of information and experience, and technology transfer. In this regard, research institutions with specific expertise should undertake twinning arrangements with other institutions to enhance such collaborations.

13.3 For Commonwealth Heads of Government Meeting

- Recognising that international attention is now focused on the role of the world's forests in global climate, environmental amelioration, biodiversity and the livelihoods of rural people; and
- following the 1992 UNCED meeting in Rio de Janeiro and the international initiatives such as the UN Commission on Sustainable Development (CSD) which are now in place; and
- being cognisant of the long history of Commonwealth forestry and the unique position of the Commonwealth Forestry Conference in bringing together foresters

from all ecological zones, government and non-government institutions, and representatives from production, processing and the timber trade;

the fourteenth Commonwealth Forestry Conference recommends:

- that Commonwealth Ministers responsible for forestry should meet in the first half of 1994 to give the necessary leadership in Commonwealth countries' preparation for the UN Commission on Sustainable Development's 1995 review of progress on forestry.
- that countries should continue to give full support to Commonwealth organisations and agencies involved in forestry, and field initiatives.

14 Fifteenth Commonwealth Forestry Conference, Victoria Falls, Zimbabwe 12th–17th May 1997 [17]

Some 194 delegates from countries other than Zimbabwe and from International Organisations and Agencies attended, plus over 100 from Zimbabwe itself. In all 30 Commonwealth countries and agencies were represented.

14.1 Opening Address

In his opening address, the President of the Republic of Zimbabwe, Comrade R.G. Mugabe, defined the objectives of the Conference:

- discuss and prepare agreed recommendations on forestry and related issues for the Commonwealth Heads of State and Government meetings; and,
- promote co-operation among members of the Commonwealth in various fields of forestry.

He noted that the theme of the Conference was "Forestry in a changing political environment: Challenges for the 21st Century".

He said that the theme of this, the third conference to be held in Africa since 1920, "could not have come at a more appropriate time than now, when more than 11 million hectares[3] of the world's tropical forests and woodlands are being destroyed annually. The major reason for the deforestation in developing countries has been the increasing dependency on natural resources, particularly agriculture, land, fuelwood, construction poles, commercial timber and other non-timber forest products"

Increasing population in developing countries has not been matched by economic growth (as it has in developed countries) but by agricultural expansion (on forest land) and expanding household uses of wood, with increasing poverty and worsening

[3] "more than 11 million hectares" has been quantified as 15 million hectares, and that is the figure that is now widely accepted.

environmental degradation. He then went on to stress Zimbabwe's commitment to reducing the adverse impact of the exploitation of forests and other natural resources by such measures as:

- helping local communities to mobilise themselves into effective forest management units practising and committed to sustainable forestry,
- by launching (in 1982) a Rural Afforestation Programme,
- by introducing alternative energy sources (e.g. solar power) to rural communities,
- by trying to introduce more energy efficient wood burning stoves into local rural communities, with limited success so far because they have not been affordable by economically marginalised communities".

14.2 Keynote Speech by Chief Emeka Anyaoku, The Secretary General of the Commonwealth

Referring to the many international meetings that have taken place in recent years, the Secretary General suggested that common themes emerging from these meetings recognise governments' efforts have often failed to stem the tide of forest loss of 15 million hectares of tropical forest falling to the axe, the torch, the bulldozer or the chainsaw every year.

"If I was to try to isolate a few of these themes they would include:

- The idea that it is in the public interest, i.e., in the interest of the well-being of us all who inhabit this planet, to manage forests sustainably and equitably;
- a recognition that to do this we need to engage civil society and ensure that people are drawn into decision-making at all appropriate levels;
- a realisation that this can only be done with political commitment on the part of governments supported by all stakeholders including the professional forestry community; and
- an acceptance by many of appropriate mechanisms for ensuring global payments for the ecological services rendered by forests to our global community.

However, the Secretary General concluded that "while there are abundant reasons for concern (on the extent to which current deforestation and degradation constitutes a threat to the maintenance of a habitable biosphere and even to planetary survival) there is no excuse for doom and gloom either. There are a number of international initiatives that have captured the imagination of many and Commonwealth countries are leading the way".

He gives examples from widely different Commonwealth countries—in Canada the dialogue between local communities and logging companies that avoided local conflict and led to the establishment of the pioneering Canadian Model Forest Programme, which in turn led to the adoption of similar model forests in Malaysia and Mexico (not of course a member of the Commonwealth).

In Africa self-help groups with help from non-government organisations have planted trees on local land and supported and organised rural development projects. Zimbabwe's forest research institutions have enabled that country to export tree seed world wide. Partnerships have been built with private ventures in sustainable timber production and sustainable wildlife management and in the export of quality timber (again from sustainable forests).

The main Conference Recommendation to Commonwealth Heads of Government was (in summary) that they should give full political commitment through participation in international forestry processes to sustainable forest management; should give full support to partnerships between communities, the state and the private sector which result in the more active involvement of local people; should support Commonwealth initiatives which are seeking to demonstrate practical approaches to sustainable forest management.

Recommendations to nations (Governments and civil society including non-governmental organisations) included a strong plea that all sections of society and in particular local interest groups be involved in decision making and subsequently in management of forest resources seen in the context of integrated land use planning; that plantation forestry be expanded on degraded land to produce timber; that political support be reflected in higher investment in sustainable forest management; that research, training and information be developed across the whole field of forestry.

Recommendations to the international community (e.g. UN bodies, Commonwealth Agencies) requested a strengthening of their role in developing partnership arrangements and exchanges of information; and that education, training and research be expanded to meet the new approaches needed in managing forests on a sustainable basis.

The Conference also recommended to the Standing Committee on Commonwealth Forestry that its role, including its information role, be widened.

References

The reports (proceedings, resolutions and recommendations) of the Fifteen British Empire/British Commonwealth/Commonwealth Forestry Conferences are held by the Standing Committee on Commonwealth Forestry, c/o the Forestry Commission, 231 Corstorphine Road, Edinburgh EH12 7AT. The titles of the reports are:

1. The Forestry Commission, London.
   British Empire Forestry Conference, London 1920.
   Proceedings, Resolutions and Summaries of Statements. HMSO 1921.
2. *Second British Empire Forestry Conference.*
   Held in Canada, July 25th to September 7th 1923.
   Proceedings and Resolutions, Ottawa 1923.
3. *Third British Empire Forestry Conference Australia and New Zealand, 1928.*
   Summary Report, Resolutions and Reports of the Committees.
   By authority H.J. Green, Government Printer, Canberra.

4. Union of South Africa.
Fourth British Empire Forestry Conference, South Africa, 1935.
Summary Report, Resolutions and Reports of Committees.
Government Printer, Pretoria, 1935.
5. H.M. Forestry Commission.
Fifth British Empire Forestry Conference Great Britain, 1947.
Summary Report, Resolutions and Reports of Committees.
London, HMSO 1947.
6. Forestry Commission. Empire *Forests and the War*[4].
Statistics Prepared for the Fifth British Empire Forestry Conference, London HMSO 1949.
7. *Sixth British Commonwealth Forestry Conference Canada, 1952.*
Proceedings Summary Report, Resolutions, Reports of Committees.
Government Printer, Ottawa 1953.
8. *Seventh British Commonwealth Forestry Conference Australia and New Zealand 1958.*
Proceedings Summary Report, Resolutions, Reports of Committees.
Government Printer, Canberra 1958.
9. *Eighth British Commonwealth Forestry Conference, East Africa 1962*
Proceedings, Committee Reports and Resolutions of the Conference.
Government Printers Kenya.
10. *Ninth Commonwealth Forestry Conference, India 1968.*
Proceedings, Committee Reports and Resolutions. Available from the Standing Committee CFC, c/o Forestry Commission, Edinburgh, UK
11. *Tenth Commonwealth Forestry Conference, UK 1974.*
Proceedings, Committee Reports and Recommendations.
Edinburgh, Forestry Commission 1975.
12. Frank Fraser Darling, *The Forest and Global Environment.* Forestry Commission, London, 1974.
13. *Report of the Eleventh Commonwealth Forestry Conference, Republic of Trinidad and Tobago, 1980.*
Proceedings, Committee Reports and Recommendations.
Forestry Division of the Ministry of Agriculture, Lands and Fisheries, 1981.
14. *Proceedings of the Twelfth Forestry Conference*, edited by
S.G. Glover, J.T. Arnott and C.E. Brown.
Victoria, Canada. September 8th–22nd, 1985.
Pacific Forestry Centre, Victoria, British Columbia, Canada.
Ministry of Supply and Services, Canada 1986.
15. *Forestry: A Multiple-Use Enterprise.*
Proceedings of the Thirteenth Commonwealth Forestry Conference.
Rotorua, New Zealand, 1989. Available from the Standing Committee CFC, c/o Forestry Commission, Edinburgh, UK
16. *People, The Environment and Forestry—Conflict or Harmony.*
Proceedings of the Fourteenth Commonwealth Forestry Conference,
Kuala Lumpur, Malaysia 1993.
Ministry of Primary Industries, Malaysia.

[4]This report covers all the countries that were part of the British Empire during the 1939–1945 World War.

17. *Forestry in a changing political environment. Challenges for the 21st Century. Proceedings of the Fifteenth Commonwealth Forestry Conference held in Victoria Falls, Zimbabwe, 12th–17th May 1997.* Zimbabwe Forestry Commission, Harare, Zimbabwe, ISBN 0–7974–1809–6.

Chapter 13

Comments and Prospects

1 Author's Comments on British Forestry

1.1 1919–1969

Establishing a reserve of growing timber

Britain lacked a national forest policy until 1919, when the Forestry Commission was set up, with a mandate to establish a reserve of growing timber. The restatement of that policy in the immediate post (1939–1945) war years—came when several important factors favoured the implementation of that policy.

Leadership

In the years (1919–1939) between the two world wars when governments of the day wavered in their support of the forest policies they had so enthusiastically embraced in 1919, R.L. Robinson (later Lord Robinson), the Technical Commissioner from 1919–1932, fourth Chairman 1932–1952, fought publicly and forcefully for the Commission to be funded to do the job for which it had been created in 1919. When the large afforestation programmes of the post (1939–1945) war years—which he had so ably advocated—were accepted by Government, field staff at all levels were made aware that they were embarking on a task of national importance. To this end, Lord Robinson travelled extensively to all parts of the country to meet and talk to field staff and to inspect their work at first hand. He also met influential private woodland owners on these visits.

Due in large part to this leadership from the front, morale was high; people were proud to work for the Forestry Commission.

Availability of Land for Planting

There were available on the open market very large areas of land too poor to support any form of agriculture other than sheep grazing. And as Lord Lovat commented in his speech to the first Commonwealth Forestry Conference in 1920, some of this land was capable of carrying only one sheep per four or five acres (2 hectares). Unlike the situation in much of Continental Europe and the Nordic countries, large areas of this land could often be purchased from a single owner in large parcels or leased on long lets at reasonable—some might say nominal—rents.

Afforestation Techniques on Poor Land

Techniques for establishing plantations of Sitka spruce and other exotic conifers on poor upland soils—often peat—had been tried out by one or two private estates. Private planting on other site types suitable for native and other exotic species had also been carried out—on a much larger scale—by some private forest owners. The Forestry Commission had a base—albeit a small one—on which to build up its knowledge of how to afforest a range of soils, particularly those unsuited to agriculture.

Private Woodland Owners Co-operation

Private woodlands had been exploited for the war efforts in 1914–1918 and 1939–1946 and many estate owners were keen to restock their woods either by their own staff or by lease to the Forestry Commission. In the 1960s private forestry investment companies were formed to create and manage new plantations—often on a large scale—on behalf of wealthy private individuals who could take advantage of the then favourable tax regimes that forestry offered.

Research

Success breeds success and the Forestry Commission's good performance in achieving the afforestation of large areas of its own land and persuading the private sector, through grants, to play its part in establishing reserves of standing trees, meant that it was able to obtain substantial funds for research into and development of improved plantation establishment techniques.

Although British foresters were taught and in some cases private woodland owners practised what is now knows as "continuous cover forestry", the most common system in the management of coniferous woodlands was to plant an area, thin the crop to get an intermediate yield and to fell at economic maturity. In the 1950s Britain became one of the leaders in Europe in this method of establishing and managing coniferous woodlands, especially on land which had not previously been wooded. A wide range of aids to woodland management was developed by forest research, comprising yield and other tables which modelled the growth of all the commonly planted species (of broadleaves as well as conifers) under different site conditions and under different thinning regimes. The main aim of British forest management was to maximise wood production as required by Government and it was only after a period of some 45 years of large scale afforestation that the Government (in 1963) directed the Commission to give increased emphasis to the provision of public access and recreation and to amenity and the protection of wildlife. This was a policy directive that the Commissioners were only too pleased to be given and inter alia they promptly appointed a specialist officer to liaise with the Nature Conservancy Council to improve nature conservation throughout the Commission. They also appointed a leading landscape architect of international status, Dame Sylvia Crowe, to advise on the layout of plantations and the conduct of all forest operations with an impact on the landscape [**1**].

One advantage that Britain claimed over many European countries was that by placing forest research firmly under the Forestry Commission, the Commission could ensure that the research programme gave high priority to serving the practical needs of field managers—in both the Commission and the private sector; some so-called "basic" or "fundamental" research was undertaken either directly or under contract by Universities and other appropriate institutions.

Housing for Supervisors and Forest Workers

The Forestry Commissioners had good personal relationships with senior politicians and were able to persuade the Government to embark on housing schemes in rural areas to house the forest workers and the supervisors urgently needed for the expansion of the forest estate. Many ex-service men setting up home for the first time were attracted to forestry and the chance to live in the country rather than in industrial cities and towns.

Public Recreation

In acquiring land for planting, the Commission often had to acquire substantial areas that would not support tree growth. Whilst the Treasury understandably discouraged the acquisition of such unproductive land even at nominal prices, many Labour (socialist) politicians agreed with the Commission that such land had value for the nation as recreational land to be thrown open to the public for "fresh air and exercise". In general private landowners felt that recreational opportunities on Commission (i.e., State) land reduced the pressure on privately owned land. Unplantable, open hill land became the core of the early National Forest Parks. In time, the whole Commission estate, woodlands as well as unplantable areas was declared "open to the public for fresh air and exercise". The only exceptions were due to restrictions on leased land (restrictions now actively being bought out by the Commission). For many of Britain's largely urban population this was seen as the main benefit of forestry.

The relatively new facilities for which the Forestry Commission made a charge, such as camp and caravan sites, were also popular with the general public.

In 1965, at an open meeting on a proposed caravan site in Scotland, a private individual challenged the Forestry Commission's powers under the Forestry Acts to spend public money on creating such facilities. This view was upheld for Scotland by the Lord Advocate, Scotland's senior law officer, and for England and Wales by the Lord Chancellor.

After discussion with the Forestry Comissioners and the Forestry ministers, the government decided to insert clauses into the Countryside Acts for Scotland (1967) and for England and Wales (1968). Theses gave the Forestry Commission specific powers to provide facilities for recreation and to manage trees in the interest of amenitity—as indeed the Commission had been doing for some thirty years. No one seemed to notice and one wonders whether the individual was ever aware of the furore he had created.

One should record that the particular project which sparked off the affair was never undertaken (J.W.L. Zehetmayr, personal communication).

Technical Co-operation with Private Sector

At local level a good spirit of co-operation between Forestry Commission staff and those on private estates facing similar problems of expanding their forestry activities was the rule rather than the exception. And this was true at regional and national level—without the private woodland owners associations losing their ability to fight their corner when they felt they were being unfairly treated by the Forestry Commission or by the Government (e.g. over tax matters).

International Co-operation

Although new to the European forestry scene, the Forestry Commission played an active role in the setting up of the European Forestry Commission (EFC) of the Food and Agriculture Organisation (FAO) and its subsidiary bodies and in turn kept the (British) private sector informed about the wider, European, forestry scene. A great deal of useful discussion took place outside the formal meetings, often on the field trips organised by member countries. Particularly influential was the Joint EFC/FAO/ILO Committee on Forest Working Techniques[1] which in its early years was concerned with improving all aspects of operational efficiency in the harvesting of timber, training, health and safety of forest workers and accident prevention. Its influence led directly to the setting up of the Forestry Commission's Work Study Branch, which was able to draw on expertise from countries with similar conditions to those in Britain, notably the Nordic countries and to make its findings available to private owners and timber merchants that bought timber standing from the Commission and private owners.

---

[1]Whose members included the countries of Europe, including Scandinavia, the USSR, Canada and the USA. Without the major improvements in harvesting timber that resulted from the Work Study Branch's activities, it would have been impossible to supply the relatively large quantities of pulpwood and panel products material required by the new industries being established in the 1960s.

Marketing of British Timber

The year 1963 marked the announcement of an important policy change by the Commissioners. The marketing of timber was to be regarded as of equal importance to the establishment of a reserve of growing timber through the large programme of new planting and re-stocking of old woodlands; the implication was that this change would apply to the private as well as the state sector. In the 1950s the Commissioners had in fact foreseen the need to find markets for the increasing volumes of timber that could and should be harvested from the new plantations they and the private sector had been establishing at an ever increasing rate, and had appointed an advisory committee, with wide ranging expertise, to advise them on what measures were necessary to find adequate markets for this timber. A small specialised research and development group (the Utilisation Development Section) was set up at Forestry Commission Headquarters working directly under the Director of Research. Contrary to the impression given ("he knew that marketing would not become a top-level matter of policy in his time") in George Ryle's "Forest Service" [**2**] Lord Robinson personally involved himself in the setting up of the Committee, in its deliberations and the programme of work of the Utilisation Section.

Prior to his departure to Canada in 1952, to attend the Sixth Commonwealth Forestry Conference, he suggested to the writer (then head of the Utilisation Development Section) various ideas that might be pursued and directed that a detailed programme and budget be prepared for discussion on his return. He specifically mentioned the building of an experimental sawmill in West Scotland which had relatively large quantities of Sitka spruce coming to sawmill size.

Although in the Commission generally, economy measures were very much the order of the day, he indicated that imaginative and bold initiatives would be fully considered even if they required substantial funding. Sadly Lord Robinson died in Canada; the development work he initiated continued for many years!

1.2 1970–1980

The Dedication Scheme (Basis II) which became operational in 1948, was the hugely successful grant-aid scheme for encouraging private owners to restock woodland which had been felled or creamed off to meet war-time needs and to engage in new planting (afforestation). In return for grant-aid owners had to covenant not to use the land dedicated to forestry for other purposes and to bind their heirs and successors to observe this restriction. One reason for this was the Commission's desire to help build up forest estates with, in due course of time, multi-purpose woodlands sitting comfortably in the landscape and in harmony with local wildlife, rather than just wood-producing plantations. Even those whose immediate personal objective was to take advantage of the friendly forestry tax regime would under

Dedication Basis II have to manage their woods—in perpetuity—under plans agreed with and approved by the forest authority. These plans would include amenity and nature conservation considerations. It was ironic therefore that the Government should close Dedication Basis II in 1972 to new applicants[2] and substitute in 1974 [**3**] a Basis III Scheme that allowed an owner to opt out of Dedication at the end of a rotation and use the land for purposes other than forestry.

One explanation is that there was a small but influential school of thought within the civil service that Dedication Basis II with its "in perpetuity" obligation on the Government's part was too generous, given forestry's favourable tax regime, and that in any case it had served one of its purposes—to give some compensation to private owners for the compulsory war-time fellings at strictly controlled prices that had often creamed off the best trees in their woodlands. However, the new Basis III Dedication Scheme of 1973/74 continued to give grant aid and the extra conditions imposed in respect of public access and consultation with agricultural authorities and environmental interests proved not to cause the private sector any difficulties.

The Forestry Commission's The First 75 Years [**4**] notes that "over a period of 10 years four grant schemes were closed and three were introduced". Given their predecessors track record in earlier difficult times, one wonders how the Forestry Commissioners of the day tolerated this state of affairs, without any public sign of disquiet. In similar vein the Commissioners were strangely silent when in 1973 quite out of the blue the Government's forest policy statement of October 1973 emphasised the need for forestry to encourage and secure effective integration between forestry and agriculture. It is not clear what was meant by integration with agriculture nor from where the concept emanated. In a very few years "effective integration with agriculture" as a condition of grant aid to private forestry was quietly dropped and has not re-appeared. However, the decade ended on an up-beat note when in 1979 the influential Countryside Review Committee (with members from every Government organisation with any connection with the countryside) published a report which *inter alia* said:

> The Countryside Review Committee does not believe that controls (planning controls over agriculture and forestry) would be justified or would serve a useful purpose.

It was interesting, not to say instructive, to observe during the meetings of the Committee, the reactions of some members—whose knowledge of the countryside was in reality very limited—when taken to visit some of the "man-made conifer slums" of the popular press and to meet and talk to foresters managing what turned out to be pleasant varied woodlands with abundant wildlife that produced timber

---

[2]Though those in the scheme could continue, as was their prerogative, the grant levels were frozen so that they became unattractive compared to new schemes.

for a range of forest products industries and at the same time provided recreation for large numbers of people, many of whom lived in towns and cities.

1.3 1981–2000

The sale of land and plantations by the Forestry Commission, forced on it by the Government under the highly unpopular 1981 Forestry Act, caused some misgivings about the Government's plans for the future of state forestry.

The Chancellor of the Exchequer's March 1988 Budget announcement of fundamental changes to the long established forestry tax regime "the engine that had driven much private planting was stilled at a stroke" [**4**]—indicated that it was not only state forestry but the long-term forest policy hitherto supported by all shades of political opinion that was at risk.

When the 1993–1994 Forestry Review was set in motion by Government, it was widely believed that the intention was to make a case for the sale of all the Forestry Commission woodlands and other assets to the private sector, retaining in the public domain only the Commission's forest authority role (e.g. management of grant aid to the private sector; licensing of all tree felling; plant health in the field of forestry).

In a very short time forestry and forestry industry organisations opposed to the privatisation of the Commission started work on making the case against it. In addition, and probably for the first time, individuals with experience in forestry or the forest products industries formed groups to articulate, in co-operation with other groups, the case against privatisation. Former Commission staff played an important role in these groups, which also received guidance from senior politicians as to how to present their submissions to Government to best effect.

The following are some of the points which were compiled by foresters and others to act as an aide-memoire to individuals and organisations arguing the case against privatisation:

The Case for Privatisation is Weak

- Short-term net revenues from sale would be small.
- Long-term costs to the taxpayer would be increased unless present grants were discontinued and that would halt afforestation.
- There is no state monopoly to be overthrown.
- There is no evidence that greater efficiency would ensue.

Freedom of the Forests would be Lost to the Public

- National legislation to preserve freedom (of access by the public) would ruin sale prospects, and risk loss of privacy in the present private sector.
- Britain would become unique in the developed world in having no extensive forests open to everyone.

Forestry Development would be Arrested

- Multi-function forestry management and silviculture would be subordinated to profit-taking, probably by multi-national industries. Wildlife conservation, some public recreation facilities, and some landscapes, would suffer in time.
- The public voice in forestry development would be reduced.
- The marketing, and to some extent the processing of home-grown wood would be adversely affected.
- The prospect of a self-financing public Forest Enterprise by 1996 would be lost.
- The Forest Authority, including its successful R & D arm, would almost certainly be quickly absorbed by one or more of the large Ministries, such as Agriculture or Environment.

Under the headline "Reason wins over political dogma" the Scotsman's editorial of 11-04-94 [**5**] includes the following:

"The Government's drive to privatise simply for the sake of it—a legacy of the Thatcher years—thankfully appears to be grinding to a halt. The news that ministers will shelve some of the most controversial plans, including the sale of the Forestry Commission, is belated recognition of the widespread opposition to some of their proposals.

The news is particularly welcome in relation to the Forestry Commission, whose future as a public body under the Government's review of forestry has not looked too bright. Although the Scottish Secretary, Ian Lang, and the Scottish Office minister responsible for forestry, Sir Hector Monro, have never been in favour of the plans, they had the Environment Department and the Treasury lined up against them. The Treasury in particular, viewing matters from an economic perspective, probably saw the privatisation of the Commission's commercial division, Forest Enterprise, as a nice little earner with a potential asking price of about £1 billion. Now, however, the fear of further political disasters in the wake of fresh privatisation schemes has clearly alarmed an accident-prone Government and helped save the day for the Commission. Indeed, the debate has never been one simply of economics. The way in which forestry is run in Scotland—and in the rest of Britain for that matter—is of crucial importance for the future of the environment and the sustainability of the countryside.

Forestry Commission forests are public places and there is in general no problem with access to its vast holdings. Privatisation carried with it the risk of greatly reduced public access and the real fear that private operators would not have been able to guarantee the regular supply of timber, regardless of market price, maintained by Forest Enterprise. Now the Government, instead of selling it off, envisages more modest plans to expand its commercial activities, which seems a much more sensible approach.

The Government had a formidable and diverse body of opinion lined up against it over the privatisation of the Forestry Commission. Apart from the more obvious opposition of environmentalists, opponents ranged from private timber growers and wood processors to the Scottish Landowners' Federation and the Royal Society for the Protection of Birds. Mr. Major's decision to listen rather than push ahead regardless of the objections, is to be welcomed. As we have pointed out before, much could be gained by putting some forestry land into the hands of local communities, but a wholesale sell-off would have been political dogma for its own sake. The decision to shelve the privatisation plans is a victory for reason and common sense."

An example of the "formidable and diverse body of opinion lined up against the Government" is the list of signatories to an open letter to the Prime Minister dated 9th March 1994.

1.4 An Open Letter To The Prime Minister

Future of the Forestry Commission—9th March 1994

From recent media coverage of the Forestry Review Group, it appears that the prospect of privatising the Forestry Commission has receded and that the Government will be willing to consider other options, as Ministers have consistently claimed that they would. This is to be welcomed. The notion of privatisation has been met with widespread hostility. The problems identified by the forestry industry, the wood processors, environmentalists and the public who enjoy the forests are clearly proving insurmountable.

Unfortunately, the alternative options reported in the press have little to commend them and there is a risk of their diverting attention from an altogether more positive solution. We, the undersigned, urge you to use your review to focus Britain's forest service on the needs and opportunities of the new century. Whereas most changes would impose initial and ongoing costs, those proposed below can be achieved at little cost to the tax payer.

We suggest that the Commission should be charged by Ministers to:

- safeguard public access to their woodlands and, where possible, extend it by establishing new woodlands;
- create new woodlands close to towns and cities for people to use for relaxation and recreation;
- have a special duty to look after our heritage of native woodlands and genepools of native species, and to extend native woodlands wherever possible, such as the Caledonian pine forest;
- care for woodland biodiversity, including particular care for those species that are rare or threatened, such as red squirrel, pine marten and nightjar;
- be a reliable, long-term supplier of wood for industry, and ensure that the potential for increased wood production, processing and jobs in Britain is fully realised without risk of long term site or soil degeneration;
- diversify and landscape the older, extensive and uniform coniferous plantations;

- take a lead in expanding Britain's forests for public recreation, wildlife habitat, wood production and landscape and foster research and development to that end. Promote and interpret forestry matters to the public.

We believe that the Government should allow the Commission to use any surplus from timber sales and the proceeds from land disposals and commercial recreation to achieve these objectives. The facility to purchase as well as sell land should replace the current sales-only programme which is reducing the area of publicly accessible woodland.

We also believe that Ministers should require the Commission to encourage greater participation by others in its affairs both by commercial partnerships, and by more non-commercial partnerships with environmental and wildlife organisations as well as local communities.

It is important that the Government continue to use the Commission to support private forestry through a well designed system of grants. Increased private investment in extending productive woodland is clearly much more in the nation's interest than any trading in woodlands which already exist.

The recent reorganisation of the Commission has been widely welcomed and is proving well able to balance the needs of the wood-using industries, environmental interests and the public. Any further subdivision of the Commission, in particular along functional or national lines would be costly and inefficient; and it would disrupt progress towards meeting the Government's forestry objectives.

The Forestry Review has revealed the widespread trust and goodwill which many organisations and individuals have for the Commission as an agent of Government. These are invaluable assets and we urge you to make full use of them by creative plans for the Commission's future. We believe that this is the only route forward that will effectively and efficiently deliver Government forestry policy and properly fulfil the Government's promises to safeguard public access and their commitment to forest sustainability as expressed at Rio de Janeiro and, more recently, at Helsinki.

We are,
Yours faithfully
Signatories as listed below:

His Grace the Duke of Somerset, President, the Royal Forestry Society
Lord Taylor of Gryfe, Chairman, All Party Parliamentary Forestry Group
Sir Michael Atiyah, OM., PRS., Trinity College, Cambridge
Sir David Montgomery Bt., former Chairman of the Forestry Commission
W.B.Banks, Professor of Forest Sciences, University of Wales, Bangor
David Bellamy, Botanist
Dr. J. Burley CBE., Director, Oxford Forestry Institute, University of Oxford
Tim Cordy, Chief Executive, Royal Society for Nature Conservation
Phil Drabble, Naturalist

Gwyn Francis, former Director General, Forestry Commission
David Goss, President, Royal Scottish Forestry Society
George Holmes, former Director General, Forestry Commission
F.C. Hummel, former Head Forestry Division of the Commission of the European Communities
Klaus Kohler, Managing Director, Caberboard plc.
Roy Lancaster, VMH, Horticulturalist
Roy MacIver, Secretary General, Convention of Scottish Local Authorities
Dr. D.C. Malcolm, Convener of the School of Forestry, University of Edinburgh
Aubrey Mattingly, Chief Executive, Ramblers Association
H.G. Miller, Professor of Forestry, University of Aberdeen
Jonathan Porritt, Environmentalist
David Seal, former Chief Research Officer, Forestry Commission
Barbara Young, Chief Executive, Royal Society for the Protection of Birds.

The Government's move from long-term to short-term grant-aid schemes for private forestry reversed the earlier policy of encouraging owners and their successors to maintain in perpetuity the woodlands they established. Now the Government proclaims that woodlands should be managed on a sustained yield basis (i.e. a sustained yield of all the goods and services that they can provide) but seems no longer so willing to produce the financial and other incentives that encourage and enable the private sector to meet this target. It is to be hoped that the (1999) announcement of the introduction by the Forestry Commission of 20 year Forest Plans for the larger private forests signals a change of policy by Government.

The mechanisation of forest operations has dramatically reduced the number of forest workers employed in both Forestry Commission and private woodlands. The introduction of electronic data-processing and the relative ease of travel have also led to a reduction in the number of supervisors and of managers at all levels, including foresters resident in or near forests. This is especially true of the Forestry Commission. Whilst it is relatively easy to build up mathematical models to guide managers in the production of wood from large-scale even-aged coniferous plantations, the management of woodlands of broadleaves and conifers of a range of ages to produce not only timber but other goods and services on a sustained yield basis over a relatively long period of time, cannot so easily be done through mathematical modelling. In the first place the collection of the large amounts of essential basic data required by the computer programmes has to be done on the ground by experienced people; then the interpretation and implementation of the proposed programmes also needs to be done by experienced foresters who have to balance all the many considerations that the programmes suggest. There is a need to review the (low) levels of field managers with which the Forestry Commission now operates, especially in light of the findings of European Timber Trends and Prospects: Into the 21st Century (ETTS V) which states that:

- intensifying management, in the light of careful determination of society's wishes for what the forest should provide, could increase the contribution of the forest to the sustainable development of the European continent.

Note: Europe in the European Timber Trends and Prospects studies includes the Nordic Countries, the Baltic Countries and all other countries of Europe. The former USSR (now Russia), the United States of America, Canada and Israel are members of the UN Economic Commission for Europe and participate in its work, but unless specifically mentioned in the text of the various European studies, data on forests and timber products for these countries and Belarus and Ukraine is excluded.

2 Prospects for the Year 2000 and Beyond

At the close of the 20th century, the outlook for British forestry in the opening decades of the 21st century is different from that which influenced the UK government in its formulation of a forestry policy in 1919.

The 1919 UK forestry policy arose out of the perceived need to provide adequate supplies of timber to meet essential national requirements during a future major war. This was translated into the need to build up a strategic reserve of growing timber in the UK.

By the close of the century much more emphasis was placed on the environmental and recreational roles of the forest whilst still maintaining wherever possible and appropriate the objective of cost-effective production of wood [**6**]. In the context of the role of the forest as a producer of wood the following extract from a recent study is appropriate [**7**]. "The forest industries are among the more important industrial sectors in Europe. They are the customers for the bulk of the wood produced in Europe's forests. The sale of wood to them is the source of the major part of the revenue that forest owners derive from their forests. Even though the non-wood goods and services of the forest are of increasing importance, it is inevitable that wood production will continue to figure prominently in most management plans.

Consequently, the maintenance of a healthy forest industry is a key element in achieving the economic, social and environmental sustainability of Europe's forests."

A working paper for ETTS V, "The policy context for the development of the forest and forest industries sector in Europe" [**8**] explores those changes in major policy areas that are having or would have, over the next 20 years, significant impacts on future demand for and supply of wood and forest products in Europe.

These are:

1. Policies to stimulate economic expansion.
2. Policies to stimulate efforts to find alternative sources of energy and to conserve energy.
3. Policies to intensify the protection of the environment and to promote sustain-

able development.

4. Policies relating to rural and regional development, including changes in agriculture.

It is not possible to do justice to the discussions, arguments and conclusions presented in the working paper. But at the risk of over-simplification of complex issues, four points may be noted.

First, "countries in transition", i.e. former planned-economy countries now moving towards free market economies, include some important producers of wood in the Baltic, in Eastern Europe and that major producer of wood, the former Soviet Union (Russia). Their degree of success in managing their forest resources and forest products industries will have a significant impact on the timber trade and timber economy of Europe.

Second, although the use of wood as a source of energy is a possibility through the establishment of so-called energy plantations, through greater use of small-sized forest thinnings and logging residues from conventional high forest and by making more use of waste paper as a fuel, it would probably require a substantial rise in energy prices to justify the investments necessary to establish "energy plantations" on a large scale.

Policies to seek alternative energy sources to fossil fuels and to reduce the harmful impact on the environment (including climate change) of burning fossil fuels would benefit wood and waste paper provided that the environmentally benign nature of wood (including waste paper) was demonstrated which "should not be difficult", the authors say. In this context afforestation for fixing CO_2 could also become an important role for forestry.

Third, if support for agriculture on marginal land is reduced then support for afforestation may take its place—not always with timber production as its main objective, but also for environmental reasons (see point four below).

Fourth, the greater emphasis on the environmental and recreational role of the forest may have a negative effect on yields of wood per unit area (though this will not necessarily reduce the overall benefits or the financial yields on any given site). On the other hand, areas afforested for purely environmental reasons (e.g. near towns and in industrial areas) will almost always make some contribution to wood supply.

As noted above, there is a trend in Europe for the State to be less directly involved in owning and managing forests through a central government department of forestry. In Britain, expansion of the State forests managed by the Forestry Commission had been halted by the year 2000. But the sale of Forestry Commission woodlands to private owners, initiated by the Conservative Government in the 1980s was halted by Labour when they came to power. Private forests continued to expand through afforestation (new planting) encouraged by Government grant-aid administered by the Forestry Commission; the new schemes included the establishment of woodlands with environmental and recreational objectives taking precedence over (but not excluding) the production of timber.

The conclusions of study ETTS V [**9**] are:

"The study shows that given continuing economic growth and competivity on price and performance of forest products, European consumption and production of forest products will continue to grow steadily. Waste paper recycling will intensify and net imports grow. Nevertheless, the demand for European roundwood will increase. Europe's forest will, however, be able to meet this challenge by increasing harvest levels, while remaining within the limits of sustainable wood supply".

Whereas earlier timber trends studies (published in 1953, 1964 and 1969) emphasised the difficulty of Europe meeting its growing demand for wood and wood products from European forests and from imports, including wood from tropical forests, the earlier ETTS published in 1976 and covering the period to 2000, laid little emphasis on the "world shortage of wood" scenario. Now ETTS V [**9**] is more confident of the ability of tropical and semi-tropical countries to supply increasing amounts of wood and wood products to world markets from the very extensive plantations they have established in recent decades.

The message of ETTS V for British forestry is spelt out in the closing paragraphs.

"One consequence of the discussion about the objectives of forest management has been the tacit de-emphasis of the wood production objective in the public debate if not in the day to day priorities, and above all, the operating budgets of forest owners and managers.

However, ETTS V has confirmed that there is and will be a demand for wood from European forests and that wood can be produced on a sustainable basis for the foreseeable future in Europe, alongside the non-wood benefits of the forest".

And calling for more intensive management of Europe's forests the study adds:

"The situation described above, of changing priorities and continuing demand for wood and non-wood goods and services, represents a challenge and an opportunity for European forest policy. The *challenge* is to define and then to deliver the best possible combination of goods and services from the forest; the *opportunity* is that there is a potential, proven by ETTS V to increase wood production and a likely possibility to increase supply of other goods and services. Thus, intensifying management, in the light of careful determination of society's wishes for what the forest should provide, could increase the contribution of the forest to the sustainable development of the European continent."

References

1. Dame Sylvia Crowe. *The Landscape of Forests and Woods.* Forestry Commission Booklet 44. HMSO 1978.
2. George Ryle, *Forest Service.* David and Charles, Newton Abbot, 1969.

3. 54th CAR, p.8.
4. *The Forestry Commission. The First 75 years.* 1994
5. *Editorial*, The Scotsman, 11-04-1994.
6. F.C. Hummel and H.A. Hilmi, *Forestry policies in Europe. An analysis.* FAO Rome Forestry Department Paper 92. 1989.
7. T.J. Peck and A. Ottitsch, *The European forest as a source of industrial raw materials.* European Forest Institute Land Use Policy 17 (2000) 209–219.
8. T.J. Peck and J. Descargues, *The policy context for the development of the forest and forest industries sector in Europe.* United Nations ECE/TIM/DP/11 New York and Geneva, 1997.
9. *European Timber Trends and Prospects: Into the 21st Century* (ETTS V). Geneva Timber and Forest Study Papers. ECE/TIM/SP/11, United Nations, New York and Geneva, 1996.

Appendix A

Abbreviations and Glossary

A.1 Abbreviations and Acronyms

Area

acres	ac
hectares	ha

Volume

cubic metres	m^3
cubic feet	ft^3

Hoppus foot

In Britain from 1736 until the introduction of metrication in 1971 the volume of a piece of round timber or a standing tree was usually measured in "Hoppus" feet—after E. Hoppus (surveyor to the Corporation of London Assurance) when he first drew up his timber measuring tables[1].

The method is correctly known as the square of the quarter-girth measure, because instead of using the cross-sectional area of a round log times its length to obtain its true geometric volume, the system uses a nominal cross-sectional area times the length of the log to obtain the volume. The nominal cross-sectional area is one-quarter of the circumference (girth) of the round piece of timber, squared and the unit of volume is known as the Hoppus foot, sometimes written cubic feet (Hoppus measure).

The Hoppus system of measurement gives a volume 21.5% less than the true volume of the log using the rules of geometry (cross-sectional area × length) and

[1]The Foresters Companion. N.D.G. James. Third Edition. Basil Blackwell, London.

thus makes an allowance for the waste incurred in cutting the round log into a square baulk of timber.

In literature, prior to the advent of metrication it is not always clear that the use of cubic foot refers in fact to the cubic foot (Hoppus measure). Even in 2000 AD, many round-wood merchants use Hoppus measures for broadleaves.

Great Britain and International Committees and Organisations

DSIR	Department of Scientific and Industrial Research
EC	European Community
ECE	Economic Commission for Europe
EFC	European Forestry Commission (of FAO)
FAO	Food and Agriculture Organisation
FHTA	Federated Home Timber Merchants Association (England and Wales)
FPRL	Forest Products Research Laboratory (originally part of DSIR)
HGTAC	Home Grown Timber Advisory Committee
HTMAS	Home Timber Merchants Association of Scotland
ICF	Institute of Chartered Foresters
IUFRO	International Union of Forest Research Organisations
SWOA	Scottish Woodland Owners Association
TC	The Timber Committee of the Economic Commission for Europe
TGA, TGO	Timber Growers Association/Organisation

ECE, EFC and FAO are United Nations bodies.

A.2 Glossary

Afforestation

Internationally accepted term to describe the planting of trees on ground that has not previously carried a crop of trees. Synonymous with new planting in UK terminology.

Reforestation

Internationally accepted term for replanting (=restocking) land that has previously been woodland.

Heeling in

The temporary planting of seedlings and transplants (commonly in trenches) to prevent roots drying out prior to their being planted individually in their permanent positions in a plantation.

Hardwood(s)

The wood of broadleaved trees or broadleaved trees themselves.

Standing Timber

Trees standing in woods.

Standing Volume

Volume of standing timber i.e. volume of trees standing in woods.

Softwood(s)

The wood of coniferous trees or conifers themselves.

Sawlogs

Round timber that is destined to be sawn into planks or boards. In the UK sawlogs have a top diameter of at least 14 cm.

NB. In International statistics, especially those relating to the sale, trade in and/or use of roundwood, under pulpwood there can be reported volume of sawlogs traded or used; this occurs when prices for small-sized sawlogs are low and pulpwood prices are relatively buoyant so that pulp mills can afford to buy in small sawlogs to make-up their requirements for fibre.

Panel Products

Plywood, hardboard, insulation board, chipboard (=particle board), MDF (medium density fibreboard), OSB (oriented strand board). All these are wood-based.

Overbark measure (OB)

Overbark measure is obtained when the girth or diameter of a tree or round log is measured with the bark on and so includes the thickness of the bark. Overbark volume is thus the volume of wood and bark.

Underbark measure (UB)

When the girth or diameter is measured after the bark has been removed. The volume recorded is the volume of wood.

Appendix B

Index to Forestry Legislation

This index was prepared in August 1977 by the Forestry Commission for its own use and is intended as a guide to the provisions on forestry which have been made by Acts of Parliament or Statutory Instruments and were in force then.

The index is in two parts. Section B.1 is an alphabetical index by titles of the Acts and Statutory Instruments which contain forestry provisions; and Section B.2 is a chronological summary of the provisions up to 1977. The Forestry Commission does not intend to compile a similar list for the period after 1977.

B.1 Alphabetical List of Legislation in Force containing Forestry Provisions

Index by titles of legislation

Abbot's Wood (Dean Forest) Act 1870
Agriculture Act 1967
Agricultural and Forestry Associations Order 1974
Agriculture (Miscellaneous Provisions) Act 1972
Bedgebury Pinetum Byelaws 1969
Civic Amenities Act 1967
Coal Act 1938
Commons Registration Act 1965
Community Land Act 1975
Conservation of Wild Creatures and Wild Plants Act 1975
Countryside Act 1968
Countryside (Scotland) Act 1967
Criminal Damage Act 1971
Criminal Justice Act 1967
Criminal Justice Act 1972
Dean Forest (Amendment) Act 1861
Dean Forest (Gale Rents) Act 1819

Dean Forest Mines Act 1838
Dean Forest (Mines) Act 1871
Dean Forest (Timber) Act 1808
Deposit of Poisonous Waste (Notification of Removal or Deposit) Regulations 1972
Development Land Tax Act 1976
Development of Rural Wales Act 1976
Diseases of Animals Act 1950
Drivers' Hours (Goods Vehicles) (Exemptions) Regulations 1972
Dutch Elm Disease (Local Authorities) Order 1977
Dutch Elm Disease (Restriction on Movement of Elms) Order 1977
European Communities Act 1972
Field Monuments Act 1972
Finance Act 1975
Finance Act 1976
Finance Act 1977
Forest of Dean Byelaws 1975
Forest Reproductive Material Regulations 1977
Forestry Act 1967
Forestry Commission Byelaws 1971
Forestry Commission Byelaws 1975
Forestry (Exceptions from Restriction of Felling) Regulations 1951
Forestry (Exceptions from Restriction of Felling) Regulations 1972
Forestry (Exceptions from Restriction of Felling) Regulations 1974
Forestry (Felling of Trees) Regulations 1951
Highways Act 1959
Highways Act 1971
Highways (Miscellaneous Provisions) Act 1961
Hovercraft (Application of Enactments) Order 1972
Importation of Forest Trees (Prohibition) (Great Britain) Order 1965
Importation of Forest Trees (Prohibition) (Great Britain) (Amendment) Order 1974
Importation of Plants, Plant Produce and Potatoes (Health) (Great Britain) Order 1971
Importation of Wood and Bark (Prohibition) (Great Britain) Order 1977
Land Powers (Defence) Act 1958
Local Government Act 1972
Local Government Act 1974
Local Government (Miscellaneous Provisions) Act 1976
Local Government (Scotland) Act 1973
London Government Order 1970
Manoeuvres Act 1958
Mineral Workings Act 1951
National Parks and Access to the Countryside Act 1949
New Forest Act 1877
New Forest Act 1879
New Forest Act 1949
New Forest Act 1964
New Forest Act 1970
New Forest Byelaws 1970

New Forest (Confirmation of Byelaws) Orders 1954 to 1976
New Forest (Election of Verderers) Regulations 1955
Opencast Coal Act 1958
Opencast Coal (Requisitioned Woodlands) Regulations 1959
Plant Health Act 1967
Plant Varieties and Seeds Act 1964
Plant Varieties and Seeds Act 1964 (Commencement No 3) Order 1973
Race Relations Act 1968
Race Relations (Prescribed Public Bodies) Regulations 1968
Rent (Agriculture) Act 1976
Restrictive Trade Practices Act 1976
Roads (Scotland) Act 1970
Road Vehicles Lighting Regulations 1971
Statute Law (Repeals) Act 1969
Statute Law (Repeals) Act 1971
Superannuation Act 1972
Superannuation (Local Government and Approved Employment) Interchange Rules 1969
Town and Country Amenities Act 1974
Town and Country Planning Act 1971
Town and Country Planning General Development Order 1973
Town and Country Planning (General Development) (Scotland) Order 1975
Town and Country Planning (Inquiries Procedure) Rules 1969
Town and Country Planning (Scotland) Act 1972
Town and Country Planning (Tree Preservation Order) Regulations 1969
Town and Country Planning (Tree Preservation Order) (Amendment) and (Trees in Conservation Areas) (Exempted Cases) Regulations 1975
Town and Country Planning (Tree Preservation Order and Trees in Conservation Areas) (Scotland) Regulations 1975
Transfer of Functions (Wales) Order 1969
Transport Act 1968
Trees Act 1970
Tribunals and Inquiries Act 1971
Value Added Tax (Consolidation) Order 1974
Vehicles (Excise) Act 1971
Watermark Disease (Local Authorities) Order 1974
Westonbirt Arboretum Byelaws 1972
Wild Creatures and Forest Laws Act 1971

B.2 Chronological List of Legislation in Force Containing Forestry Provisions up to 1977

The forestry provisions in each Act of Parliament or statutory instrument are only briefly summarised. For further and full information, reference should be made to the text of the appropriate legislation. The major Forestry Act in force in 1977 was that of 1967 which consolidates all the Forestry Acts back to 1919.

Abbreviations		
art, Art	:	article
c, cap, ch	:	chapter
Pt	:	part
s	:	section
sch, Sch	:	schedule
SI	:	Statutory Instrument
SRO	:	Statutory Rules and Orders
reg	:	regulations

Dean Forest (Timber) Act 1808

48 Geo 3, cap 72

Provides for the inclosure of up to 11 000 acres in aggregate in the Forest of Dean for growing timber (s3) and for the throwing open of inclosures when the timber has grown past danger from browsing (s4).

[Amended by the Wild Creatures and Forest Laws Act 1971].

Dean Forest (Gale Rents) Act 1819

59 Geo 3, cap 86

Provides that the Gaveller may distrain for, or sue for recovery of, gale rent arrears (s7 and 8).

[Amended by the Statute Law (Repeals) Act 1971].

Dean Forest Mines Act 1838

1 and 2 Vict, cap 43

Makes provision for regulating the opening and working of mines. In particular, provides for the setting up of a register of freeminers (s14–22), for the granting of gales and of leases of quarries and the rules relating to them (s23–45 and s53–62), for the periodic review and arbitration over gale rents (s46–52), and for rules relating to grants in the enclosures (s64–68).

[Amended by the Dean Forest (Amendment) Act 1861, the Abbot's

Wood (Dean Forest) Act 1870, the Dean Forest (Mines) Act 1871 and the Statute Law (Repeals) Act 1969].

Dean Forest (Amendment) Act 1861

24 and 25 Vict, cap 40

Provides a declaration of the interest acquired by a galee (s1 and 3) and a determination of the liability to pay gale rents (s4). Modifies the Dean Forest Mines Act 1838 (qv) (s7–16). Varies the rules relating to the surrender and boundaries of gales and leases (s18–24). Also provides that forest workers' house sites in enclosures shall not be laid open (s26).

[Amended by the Dean Forest (Mines) Act 1871, the Wild Creatures and Forest Laws Act 1971 and the Statute Law (Repeals) Act 1969].

Abbot's Wood (Dean Forest) Act 1870

33 Vict, cap viii

Provides that Abbot's Wood shall remain unenclosed and that the owner shall not be entitled to any apportionment of gale rents under s67 of the Dean Forest Mines Act 1838 (s9).

Dean Forest (Mines) Act 1871

34 and 35 Vict, cap 85

Modifies s18–21 of the Dean Forest (Amendment) Act 1861 which relate to surrender of gales (s33 and 34) and modifies s49 of the Dean Forest Mines Act 1838 which relates to costs of arbitration over gale rents (s37).

New Forest Act 1877

40 and 41 Vict, ch cxxi

Provides the basis for that part of the management of the New Forest which is under the jurisdiction of the New Forest Verderers. In particular, provides for certain rights of enclosure by the Forestry Commissioners and for the preservation of ancient and ornamental woods (s6–8); defines the constitution, status and powers of the Verderers (s14, 16 and 20–24) and prescribes their powers to make byelaws (s25).

[Amended variously by the New Forest Acts 1879, 1949, 1964 and 1970].

New Forest Act 1879

42 and 43 Vict, ch cxciv

Gives the New Forest Verderers discretion to allow depasturing of animals in excess of commoners' entitlements and of animals belonging to non-commoners who own or occupy land adjacent to the Forest.

[Amended by the New Forest Act 1949].

Coal Act 1938

1 and 2 Geo 6, ch 52

Makes provision for the application to the Forest of Dean of certain parts of the Act relating to unification of coal mining royalties (s43). Provides that an Order in Council may be made to transfer Forestry Commission powers under Dean Forest enactments to the National Coal Board (s43).

New Forest Act 1949

12 and 13 Geo 6, ch 69

Makes further provision for the New Forest. In particular, revises the membership of the Court of Verderers (s1), lays down election procedure (s3 and 5–7) and extends the powers of the Verderers to make byelaws and levy marking fees (s9).
Provides for a Forestry Commission register of common rights (s4) and places certain duties on the Commissioners as to drainage and grazing (s11). Provides for timber enclosures up to 5000 acres additional to that provided for in the New Forest Act 1877 (s12) and for the regeneration of ancient and ornamental woods (s13).
Permits the Minister of Agriculture, Fisheries and Food to licence the use of land for various purposes, including car parks, subject to payment by the Commission to the Verderers of compensation for any interference with commoners' rights (s18) and permits exchanges of commonable for non-commonable land (s19).

[Amended by the New Forest Act 1964].

National Parks and access to the Countryside Act 1949

12, 13 and 14 Geo 6, ch 97

Provides for access agreements and orders enabling the public to have access to open country (including woodlands if in the countryside) in England and Wales (s59–83), for suspension of application of such orders or agreements during periods of high fire risk (s69), for the exemption from such agreements or orders of certain woodlands in specified circumstances (s79) and empowers local planning authorities in England and Wales to plant trees for amenity and, if necessary, to acquire land compulsorily for the purpose (s89). Supplementary provisions as to Crown land are prescribed (s101).

[Amended by the Countryside Act 1968]. [Equivalent provisions for Scotland are contained in the Countryside (Scotland) Act 1967].

Diseases of Animals Act 1950

14 Geo 6, ch 36

Assigns a meaning to "local authority" which is also applicable for the purposes of the Plant Health Act 1967 as far as it relates to Scotland (s59).

[Amended by the Local Government (Scotland) Act 1973].

Mineral Workings Act 1951

14 and 15 Geo 6, ch 60

Makes provision for the reclamation, cultivation, afforestation and treatment of land in England used for the working of ironstone by opencast operations. Empowers local authorities in the ironstone district to reclaim ironstone worked land for forestry (s16) and be reimbursed except where forestry grants have been paid on the land (s18). Provides that any such land subsequently disposed of to the Minister of Agriculture shall be deemed to have been acquired by him under the Forestry Act (s19). Also provides that the Commissioners' power to pay grants under section 5 of the Forestry Act 1967 shall include power to make grants for the afforestation of worked ironstone land and modifies, in respect, of worked ironstone land, the compulsory purchase procedure in the Forestry Act 1967 (s25). Financial provisions for reimbursements to the Forestry Fund from the Ironstone Restoration Fund are prescribed (s40).

Forestry (Exceptions from Restriction of Felling) Regulations 1951

SI 1951/1725

Provide for 3 exceptions to the requirement of a licence for felling trees additional to those specified in section 9 to the Forestry Act 1967 (reg 4) and for the ascertainment of the cubic content of trees, references to which appear in the same section 9 of the 1967 Act (reg 3). Apply to England, Scotland and Wales.

Forestry (Felling of Trees) Regulations 1951

SI 1951/1726

Make provision for various procedural matters under the Forestry Act 1967. In particular, prescribe the manner in which an application for a felling licence may be made (reg 3), the time and manner in which certain notices relating to the Forestry Commissioners' power to control the felling of trees may be given (regs 4 and 6–12) and prescribe the manner in which certain claims for compensation may be made (reg 5 and 12). Apply to England, Scotland and Wales.

New Forest (Confirmation of Byelaws) Orders 1954 to 1976

SI 1954/869, 1957/1804, 1962/596, 1970/1967, 1974/1462, 1976/2041

Confirm and make amendments to a code of byelaws made by the Verderers of the New Forest.

New Forest (Election of Verderers) Regulations 1955

SI 1955/1050

Determine the procedure for the nomination of candidates at elections of Verderers of the New Forest, the places where polling stations are established, and the procedure for voting, the counting of votes and the declaration of the result of an election.

Land Powers (Defence) Act 1958

6 and 7 Eliz 2, ch 30

Provides that where trees in England, Scotland and Wales might interfere with the use of airfields or navigational apparatus, a Secretary of State may order the removal or lopping of the trees and, in default, may remove or lop them himself; occupier is entitled to compensation and repayment of net expenses of felling (s10 and 11).

Opencast Coal Act 1958

6 and 7 Eliz 2, ch 69

Modifies the provisions of a tree preservation order where an authorisation has been granted for opencast coal operations (s2). Vests in the National Coal Board the ownership of timber felled during temporary occupation of land under a compulsory rights order (s10). Powers are conferred on the Secretary of State for Energy to make regulations adapting the compensation provisions relating to woodlands (sch 6 and 10). Provisions for Crown land are prescribed (s44). Applies to England, Scotland and Wales. [See the Opencast Coal (Requisitioned Woodlands) Regulations 1959].

Manoeuvres Act 1958

7 Eliz 2, ch 7

Provides that the New Forest Verderers shall be sent a draft of any Order to authorise the use of any part of the New Forest for defence manoeuvres and that the Verderers' consent is required to such use for more than 3 months in any period of 5 years (s1).

Highways Act 1959

7 and 8 Eliz 2, ch 25

Consolidates certain enactments relating to highways, streets and bridges in England and Wales. In particular, requires that planning authorities, when planning public paths, shall have due regard for the needs of forestry (s32 and 113).

Provides that a highway authority may not gather stones for road repairs from enclosed plantations or small enclosed woods and that a quarry owner's consent is required to quarry stone in enclosed plantations for bridge-building (s48).

Applies the provisions of the Forestry Act 1967 relating to tenants for life and other limited owners to agreements made by highway authorities relating to cattle-grids and bypasses (s92).

Disallows interference with enclosed land set apart as a tree nursery arising from the construction of a temporary highway (s114). Restricts the planting of trees near a carriageway (s123).

Provides for the pruning or lopping of trees damaging a highway by excluding sun and wind (s120) and for dealing with trees overhanging, or likely to fall on a road or path (s134). Enables local authorities to enter land for the purpose of ascertaining if a tree is dead, diseased, damaged or insecurely rooted (s256).

[Amended by Highways (Miscellaneous Provisions) Act 1961, the Countryside Act 1968 and the Highways Act 1971].

Opencast Coal (Requisitioned Woodlands) Regulations 1959

SI 1959/1427

Modify certain compensation provisions in the Opencast Coal Act 1958 in their application to woodlands.

Highways (Miscellaneous Provisions) Act 1961

9 and 10 Eliz 2, ch 63

Makes certain amendments to the law relating to highways, streets and bridges in England and Wales. Provides that, if it appears to a highway authority that, by reason of being dead, diseased or insecurely rooted, a tree is likely to cause danger by falling on a highway, road or footpath, the authority may require the owner or occupier to cut or fell the tree to remove the danger and provides that sections 134 and 256 of the Highways Act 1959 shall apply to such circumstances (s10).

Plant Varieties and Seeds Act 1964

1964 c 14

Enables the Minister of Agriculture, Fisheries and Food in England and Wales and the Secretary of State for Scotland in Scotland to make schemes prescribing the forest tree species for which plant breeders' rights may be granted (s1 and 3).
Enables the forestry Ministers to make seeds regulations conferring on the Forestry Commissioners powers to regulate the quality importation and marketing of forest reproductive material (s16) and empowers the Forestry Commissioners to establish an official seed testing station for such material (s29).
Makes provision for statutory warranties (s17), an Index of names of plant varieties (s20) and rights of appeal to a Plant Varieties and Seeds Tribunal (s10).

[Amended by the European Communities Act 1972].

New Forest Act 1964

1964 c 83

Makes further provision for the New Forest and amends the New Forest Acts 1877 to 1949. In particular, provides for the alteration of the perambulation and designation of land added to the Forest which is subject to rights of common of pasture (s1 and 2), with the consent of the Minister of Agriculture, Fisheries and Food for certain fencing by the Verderers of danger spots (s3) and certain enclosures by the Commissioners for camping sites and ancillary recreational facilities (s6), for the fixing of pannage time (s7) and for the creation of new ornamental woods (s10). Places on the Commissioners and the Verderers a duty to have regard to the preservation of amenity in the Forest (s15).

Commons Registration Act 1965

1965 c 64

Exempts from the provisions relating to commons registration the New and Dean Forests (s11).

Importation of Forest Trees (Prohibition) (Great Britain) Order 1965

SI 1965/2121

Prohibits the importation (except under licence) of certain kinds of forest trees into Great Britain from various countries abroad, with a view to preventing the introduction of diseases and pests injurious to those trees and also provides for the disposal of trees illegally landed.

[Amended by the Importation of Forest Trees (Prohibition) (Great Britain) (Amendment) Order 1974 and the Hovercraft (Application of Enactments) Order 1972].

Plant Health Act 1967

1967 c 8

Enables the Forestry Commissioners to control the introduction and spread of pests injurious to trees in Great Britain (s2 and 3).

[Amended by the Criminal Justice Act 1967, the European Communities Act 1972, the Local Government Acts 1972 and 1974, and the Local Government (Scotland) Act 1973].

Forestry Act 1967

1967 c 10

Consolidates the main forestry legislation 1919–1963.
In Part I (ss 1–8) deals with the constitution, general duties and powers of the Forestry Commissioners, particularly their powers to manage forestry land, make grants for afforestation, make dedication agreements and prevent damage by pests.
In Part II (ss 9–36) provides powers to control the felling of trees by means of licences and felling directions and sets out enforcement powers.
In Part III (ss 37–45) provides for the continuance of advisory committees and empowers the Minister to acquire land for forestry purposes and dispose of land not so required; provides for the Forestry Fund to be maintained and sets out what payments may be made into and out of the Fund; requires the keeping of annual accounts and the making of an annual report.
In Part IV (ss 46–51) provides powers to make byelaws, to enter on and survey land for forestry purposes and to enforce the byelaws.
The Act has been variously amended by:

Civic Amenities Act 1967
Criminal Justice Act 1967
Countryside (Scotland) Act 1967
Countryside Act 1968
London Government Order 1970 (SI 1970/211)
Trees Act 1970

Town and Country Planning Act 1971
Superannuation Act 1972
Town and Country Planning (Scotland) Act 1972
Criminal Justice Act 1972

Agriculture Act 1967

1967 c 22

Imposes a duty on the Forestry Commissioners to co-ordinate plans with any rural development board which may be set up (s46). Applies to England, Scotland and Wales.

Civic Amenities Act 1967

1967 c 69

Increases the alternative penalty under section 17(1) of the Forestry Act 1967 for illegal felling to £250 (s15).

Criminal Justice Act 1967

1967 c 80

Amends certain penalties for contravention of provisions in Orders made under the Plant Health Act 1967, and in the Forestry Act 1967.

Countryside (Scotland) Act 1967

1967 c 86

Empowers the Secretary of State to exclude woodlands in Scotland from public access (s15 and 19). Empowers the Forestry Commissioners to provide recreational facilities on land in Scotland placed at their disposal and empowers the Secretary of State to acquire land for the purpose (s58 and 59). Provides for the Forestry Commissioners to appoint wardens to act on Commission Land (s65). Places on the Forestry Commissioners the duty of having regard to the desirability of conserving the natural beauty and amenity of the countryside (s66).
[Corresponding provisions for England and Wales are contained in the National Parks and Access to the Countryside Act 1949 and the Countryside Act 1968.

Countryside Act 1968

1968 c 41

Empowers the Secretaries of State for the Environment and for Wales to exclude woodlands in England and Wales from public access (s19). Empowers the Forestry Commissioners to provide recreational facilities on land in England and Wales placed at their disposal and empowers the Minister of Agriculture and the Secretary of State for Wales to acquire land for the purpose (s23). Places on the Forestry Commissioners the duty of having regard to the desirability of conserving the natural beauty and amenity of the countryside (s11) and of avoidance of pollution (s38).
Empowers the Forestry Commissioners to undertake amenity planting in England, Scotland and Wales and empowers the forestry Ministers to acquire land for the purpose (s24).
[Corresponding provisions for Scotland are contained in the Countryside (Scotland) Act 1967].

[Amended by the Local Government Act 1974].

Race Relations Act 1968

1968 c 71

Makes provision for preserving rules restricting employment in the Forestry Commission (s27).

Transport Act 1968

1968 c 73

Exempts from certain permitted drivers' hours time spent driving in the course of operations of forestry, except driving on a road (s96) applies to England, Scotland and Wales.

Race Relations (Prescribed Public Bodies) Regulations 1968

SI 1968/1841

Prescribe the Forestry Commission as a public body to which section 27 of the Race Relations Act 1968 relates.

Town and Country Planning (Tree Preservation Order) Regulations 1969

SI 1969/17

Prescribe the form of tree preservation orders in England and Wales and make provision for unopposed tree preservation orders. In particular, require a local planning authority to send a copy of a tree preservation order and the map to the Conservator of Forests (reg

5(b)) and by virtue of paragraphs 1, 2 and 3 of the second schedule to the Form of Tree Preservation Order except from the consent of the planning authority under the Order the cutting down of any tree in accordance with a plan of operations approved by the Forestry Commissioners under the approved woodlands or any dedication scheme.

[Amended by the Town and Country Planning (Tree Preservation Order) (Amendment) and (Trees in Conservation Areas) (Excepted Cases) Regulations 1975].

Bedgebury Pinetum Byelaws 1969

SI 1969/312

Relate to Bedgebury Pinetum, car park and forest plots; made to preserve the trees prevent injury to the land and its amenities and to regulate the reasonable use of the land by the public for the purpose of exercise and recreation.

Transfer of Functions (Wales) Order 1969

SI 1969/388

Provides that the Ministers who are responsible in England and Wales for deciding under s44 of the Agriculture (Miscellaneous Provisions) Act 1968 what is, in substance, a "co-operative association" shall be, in relation to Wales the Secretary of State and, in any other case, the Minister of Agriculture and the Secretary of State jointly (Sch 2, PtII, para 7).

Superannuation (Local Government and Approved Employment) Interchange Rules 1969

SI 1969/997

Provide for payment of the transfer value between the Forestry Commission and local government employment (Rule 2(1) and Sch 1).

Town and Country Planning (Inquiries Procedure) Rules 1969

SI 1969/1092

Regulate the procedure (including the appearance of Forestry Commission staff) to be followed at public inquiries relating to tree preservation orders.

Statute Law (Repeals) Act 1969

1969 c 52

Repeals various provisions relating to mining and freeminers in the Dean (Pt VII of the Schedule).

London Government Order 1970

SI 1970/211

Modifies s40(2)(c) of the Forestry Act 1967 in relation to the compulsory purchase of land belonging to the Greater London Council (art 3(7)).

Roads (Scotland) Act 1970

Provides for dealing with trees overhanging roads in Scotland (s19).

New Forest Act 1970

1970 c 21

Makes the exercise of the Forestry Commissioners' powers under s23 of the Countryside Act 1968 subject to the agreement of the verderers (s1).
[Amended by the Local Government Act 1974].

Trees Act 1970

1970 c 43

Modifies s12 of the Forestry Act 1967 to enable conditions to be attached to a licence to fell trees on land subject to a dedication covenant where no approved working plan is in force (s2).
[Amended by the Town and Country Planning Act 1971 and the Town and Country Planning (Scotland) Act 1972].

New Forest Byelaws 1970

SI 1970/1068

Re-enact with amendments earlier byelaws for the forest. The byelaws are made to preserve the trees, to prevent injury or disfigurement of the land or its amenities and to regulate the reasonable use of the land by the public for the purpose of exercise and recreation.

Vehicles Excise Act 1971

1971 c 10

Provides for the annual rates of Excise Duty chargeable on tractors used in forestry (s1 and sch 3 Pt I).

Importation of Plants, Plant Produce and Potatoes (Health) (Great Britain) Order 1971

SI 1971/438

Lays down the conditions applying to plant imports into England, Scotland and Wales.

Road Vehicles Lighting Regulations 1971

SI 1971/694

Permit Forestry Commission vehicles used for fire fighting to carry one or more blue lamps (reg 64(1)(a)(iii)).

Forestry Commission Byelaws 1971

SI 1971/997

Relate to all the lands under the management of the Forestry Commission except the New Forest, the Forest of Dean and certain neighbouring parishes, Bedgebury Pinetum and Westonbirt Arboretum. The byelaws are made to preserve the trees, to prevent injury or disfigurement of the land or its amenities and to regulate the reasonable use of the land by the public for the purposes of exercise and recreation.

[Amended by the Forestry Commission Byelaws 1975].

Highways Act 1971

1971 c 41

Permits highway authorities in England and Wales to licence tree planting on highway verges (s43).

Wild Creatures and Forest Laws Act 1971

1971 c 47

Arbrogates the forest law in England and Wales and repeals spent statutes.

Criminal Damage Act 1971

1971 c 48

Makes it an offence in England and Wales to damage trees not growing wild (s10(1)).

Statute Law (Repeals) Act 1971

1971 c 52

Repeals various provisions relating to gale rents in the Dean (Pt IX of the Schedule).

Tribunals and Inquiries Act 1971

1971 c 62

Continues under the supervision of the Council on Tribunals reference committees appointed for the purposes of s20 or 21 of the Forestry Act 1967 (Sch 1, Pt I).

Town and Country Planning Act 1971

1971 c 78

Provides for the making of tree preservation Orders in England and Wales (s59–62), for penalties for non-compliance (s102–3) and for compensation (s174–5) in relation to such orders. Makes minor consequential amendments to the Forestry Act 1967 (sch 23) [Equivalent provisions for Scotland are contained in the Town and Country Planning (Scotland) Act 1972].

[Amended by the Town and Country Amenities Act 1974].

Forestry (Exception from Restriction of Felling) Regulations 1972

SI 1972/91

Excepts from the need for a licence the felling of any tree affected by Dutch Elm disease to such an extent that the greater part of the tree is dead.

Westonbirt Arboretum Byelaws 1972

SI 1972/303

Relate to certain lands under the management or control of the Forest Commissioners at Westonbirt Arboretum. The byelaws are made to preserve the trees, to prevent injury or disfigurement of the land or its amenities and to regulate the reasonable use of the land by the public for the purposes of exercise and recreation.

Superannuation Act 1972

1972 c 11

Repeals paragraph 11 of Schedule 1 to the Forestry Act 1967—which relates to superannuation of Forestry Commissioners—(s28(1)); and makes minor and consequential amendments to that Schedule (s29 and sch 6).

Drivers' Hours (Goods Vehicles) (Exemptions) Regulations 1972

SI 1972/573

Except from requirements as to drivers' hours the carriage of trees from any place where they have recently been felled (reg 3 and Pt II of the Schedule).

Hovercraft (Application of Enactments) Order 1972

SI 1972/971

Modifies the Importation of Forest Trees (Prohibition) (Great Britain) Order 1965 in relation to hovercraft or activities or places connected with hovercraft (art 5 and Schedule 2 Part B).

Deposit of Poisonous Waste (Notification of Removal or Deposit) Regulations 1972

SI 1972/1017

Exempt from the need to notify local and river authorities of the removal or deposit of waste derived from a substance acquired for application to woodlands provided it is disposed of in a safe manner on such land by or on behalf of the person who acquired it (Art 4(1)(a)). Applies to England, Scotland and Wales.

Field Monuments Act 1972

1972 c 43

Empowers the Secretary of State for the Environment to make acknowledgement payments for field monuments on private woodlands (s1).

Town and Country Planning (Scotland) Act 1972

1972 c 52

Provides for the making of tree preservation orders in Scotland (s57–60), for penalties for non-compliance and enforcement (s98–9) and for compensation (s163–4) in relation to such orders [Equivalent provisions for England and Wales are contained in the Town and Country Planning Act 1971].

[Amended by the Town and Country Amenities Act 1974].

Agriculture (Miscellaneous Provisions) Act 1972

1972 c 62

Enables the forestry Ministers to make orders specifying poisons for use against grey squirrels (s19), declares that orders made under the Plant Health Act 1967 may provide for recovery of costs by local authorities in default cases (s20) and exempts from the powers of the Secretary of State in relation to certain parks and gardens etc in Scotland land the subject of byelaws made under S46 of the Forestry Act 1967 (s23(3)).

European Communities Act 1972

1972 c 68

Enables the forestry Ministers to regulate the marketing, or the importation or exportation of seeds or any related activities and empowers the Forestry Commissioners to establish and maintain an official seed testing station for silvicultural propagating material under the Plant Varieties and Seeds Act 1964 (Sch 4D); extends the powers of the Forestry Commissioners under the Plant Health Act 1967 to cover needs arising from Community obligations (Sch 4G).

Local Government Act 1972

1972 c 70

Redefines, in relation to England and Wales, the local authorities who can be required to carry into effect Orders made under the Plant Health Act 1967 (Sch 29 para 34) [Corresponding provisions for Scotland are contained in the Local Government (Scotland) Act 1973].

Criminal Justice Act 1972

1972 c 71

Amends the Forestry Act 1967 by repealing s46(6) which loses its point with the enactment of s61 of this Act which provides that any sum received by the Secretary of State under s27(1) of the Justice of the Peace Act 1949 is to be paid by him into the Consolidated Fund.

Town and Country Planning General Development Order 1973

SI 1973/31

In general, exempts from the need to obtain planning permission in England or Wales the erection of forest buildings (except houses) and works on forest land (Art 3 and Sch 1, Pt 1 Class VII). With the approval of the Secretary of State for the Environment local authorities may direct that the exemption shall not apply (Art 4) [Corresponding provisions for Scotland are contained in the Town and Country Planning (General Development) (Scotland) Order 1975].

Plant Varieties and Seeds Act 1964 (Commencement No 3) Order 1973

SI 1973/928

Brings into force sections of the Plant Varieties and Seeds Act 1964 which are concerned with the making of seeds regulations for England, Scotland and Wales.

Local Government (Scotland) Act 1973

1973 c 65

Redefines, in relation to Scotland, the local authorities who can be required to carry into effect Orders made under the Plant Health Act 1967 (S144(1)) and removes, in relation to Scotland, the Forestry Commissioners' power of direction under s5(2) and 6(3) of the Plant Health Act 1967 (S144(3)). Amends s40(2)(c)(ii) of the Forestry Act 1967 by substituting for the words "county, town" the words "regional, islands" (Sch 27 para 166) [Corresponding provisions for England and Wales are contained in the Local Government Act 1972].

Importation of Forest Trees (Prohibition) (Great Britain) (Amendment) Order 1974

SI 1974/1

Restricts the application of the Importation of Forest Trees (Prohibition) (Great Britain) Order 1965 to living plants or trees and increases the maximum penalties for offences under that Order.

Watermark Disease (Local Authorities) Order 1974

SI 1974/768

Gives powers to local authorities listed in the Schedule to the Order to take steps in connection with Watermark Disease in willows. Gives officers appointed by these authorities powers of inspection and gives authorities powers to issue notices requiring the destruction of trees and the taking of steps to prevent the spread of the disease.

Value Added Tax (Consolidation) Order 1974

SI 1974/1146

Applies the standard rate of VAT chargeable under the Finance Act 1972 to the granting of any right to fell and remove standing timber (art 4 and sch 2).

Local Government Act 1974

1974 c 7

Removes, in relation to England and Wales, Commissioners' powers of direction under s5(2) and s6(3) of the Plant Health Act 1967 (s35 and Sch 6 para 19). Repeals s4 of New Forest Act 1970 and s5 of the Countryside Act 1968 which allowed the Verderers of the New Forest to receive grants from the Countryside Commission. The Verderers are still eligible for grants however as s9 replaces the earlier provisions.

The Forestry (Exception from Restriction of Felling) Regulations 1974

SI 1974/1817

Revoke and re-enact earlier regulations made in 1959 and also add a new exception which allows the felling of trees without a licence on land which is subject to an agreement as is described in s5 of the Forestry Act 1967 but only if the agreement satisfies certain conditions and if the felling is in accordance with a plan of operations approved by the Commissioners under the agreement.

Town and Country Amenities Act 1974

1974 c 32

Makes provision for the control of development in the interests of amenity, for the protection of trees and the preservation and enhancement of conservation areas.
Section 8 adds to the provisions in section 61 of the Town and Country Planning Act 1971 (provisional tree preservation orders) requirements as to the protection of trees in conservation areas in England and

Wales. Section 9 adds to the provisions in section 59 of the Town and Country Planning (Scotland) Act 1972 a similar requirement for Scotland.
Section 10, which is not confined to conservation areas, extends the provisions in section 60 of the Town and Country Planning Act 1971 (tree preservation orders) to prohibit the uprooting and wilful damage of trees, and modifies (including an increase in) the penalties for non-compliance with tree preservation orders. Section 11 makes similar provisions for tree preservation orders in Scotland.

Agricultural and Forestry Associations Order 1974

SI 1974/1836

Excludes from the benefit of the Agricultural and Forestry Associations Act 1962 (which exempts from registration under the Restrictive Trade Practices Act 1956 certain agreements of forestry associations), agreements made between 2 or more co-operative associations, between 2 or more federals (associations of co-operative associations) or between one or more associations and one or more federals unless they are of particular kinds defined in the Order, including exclusive dealing and sole agency agreements (art 5). Also sets out the provisions which company-type co-operatives must include in their memoranda or articles of association if their agreements are to be protected (art 4).

Town and Country Planning (Tree Preservation Order) (Amendment) and (Trees in Conservation Areas) (Exempted Cases) Regulations 1975

SI 1975/148

Make amendments to the prescribed form of tree preservation order contained in the Town and Country Planning (Tree Preservation Order) Regulations 1969. At reg 2iii add to the prescribed form of order a further circumstance in which consent under the order is not required prior to the cutting down of a tree (namely the cutting down of any tree which is in accordance with a plan of operations approved by the Forestry Commission under the approved woodlands scheme or the Basis 3 scheme where the agreement is under hand). Also specify (reg 3) exempted cases in which the protection given by section 61A of the Town and Country Planning Act 1971 is not to apply; these include felling of trees under a licence granted by the Forestry Commissioners and silvicultural thinnings.

Finance Act 1975

1975 c 7

Repeals the greater part of the former estate duty provisions for forestry and replaces them with capital transfer tax. Section 36 and Schedule 9 have effect for giving relief for woodlands. The position is that trees will be included with the deceased's estate unless an election is made by his personal representatives within 2 years of death for them to be left out of account. In this case the eventual liability to tax remains but its application is deferred until the trees are disposed of—otherwise than on a death where the relief applies. The liability to tax then applies in the case of a disposal by way of a sale to the net proceeds of the sale, after deducting harvesting and restocking costs and, in the case of any other disposal (e.g., by a lifetime gift) to the value of the trees at the time of the disposal. The rate of tax will be that of the top slice of the estate and the deferment option is only available if the deceased has owned the land for 5 years and if it is managed in accordance with a plan agreed by the Forestry Commissioners under a dedication covenant or agreement.

[Amended by the Finance Acts 1976 and 1977].

Town and Country Planning (General Development) (Scotland) Order 1975

SI 1975/679 (s107)

In general, exempts from the need to obtain planning permission the erection in Scotland of forest buildings (except houses) and works on forest land (art 3 and sch 1, pt 1, class VI). With the approval of the Secretary of State, local authorities may direct that the exemption shall not apply (art 4). Equivalent provisions for England and Wales are contained in the Town and Country Planning (General Development) Order 1973.

Forest of Dean Byelaws 1975

SI 1975/918

Revoke earlier byelaws made for the Forest of Dean by the Forestry Commissioners in 1929 and re-enact them with amendments and additions. Apply to lands in the statutory Forest of Dean and certain neighbouring areas under the management or control of the Commissioners and to which the public have access. The byelaws are made to preserve the trees, to prevent injury or disfigurement of the land or its amenities and to regulate the reasonable use of the land by the public for the purposes of exercise and recreation.

Forestry Commission Byelaws 1975

SI 1975/919

Amend the Forestry Commission Byelaws 1971 by extending exceptions to those byelaws to include certain lands in the vicinity of the Forest of Dean and also revoke a spent repeal provision in the 1971 Byelaws.

Town and Country Planning (Tree Preservation Order and Trees in Conservation Areas) (Scotland) Regulations 1975

SI 1975/1204 (S183)

Supersede the Town and Country Planning (Tree Preservation Order) (Scotland) Regulations 1948 by prescribing the form of tree preservation orders in Scotland and by making provision for unopposed tree preservation orders. The principal changes are the making of provision for the confirmation by Planning Authorities of unopposed Orders, the prescribing of exemptions from the general prohibition against the felling etc of trees in conservation areas in section 59A of the Town and Country Planning (Scotland) Act 1972, and the prescribing for the first time of the Form of Tree Preservation Order to be used by planning authorities. In particular requires a planning authority to send a copy of a tree preservation order and the map to the Conservator of Forests (reg 5(c)) and prescribes an exempted case in which the protection given by section 59A of the 1972 Act is not to apply, namely the felling of trees under a licence granted by the Forestry Commissioners and silvicultural thinnings (reg 11(ii) and (iv)); and by virtue of paragraph 1 in the Second Schedule to the prescribed Form of Tree Preservation Order except from the consent of the planning authority under the Order the cutting down of any tree in accordance with a plan of operations approved by the Forestry Commissioners under the approved woodlands or any dedication scheme.

Conservation of Wild Creatures and Wild Plants Act 1975

1975 c 48

Provides for the protection and conservation of wild creatures and of plants growing wild. Makes it an offence for any unauthorised person without reasonable excuse to uproot any wild plant but provides a defence for operations carried out in accordance with good forestry practice (sections 4 and 5). Applies to England, Scotland and Wales.

Community Land Act 1975

1975 c 77

Enables local authorities in England, Scotland and Wales to acquire, manage and deal with land suitable for development and makes provision for and in connection with public ownership of land. By virtue of section 3 and Schedule 1, all development connected with forestry, apart from the erection of dwellings, is exempted from the definition of "relevant development" and outside the acquisition power of local authorities.

Development Land Tax Act 1976

1976 c 24

Imposes in England, Scotland and Wales a new tax on the realisation of the development value of land. By virtue of S7(7)(a)(i) and (b) and Sch 4 para 5(1)(c) the carrying out of forestry building, works and operations will not rank as material development and is therefore outside the scope of the act.

Restrictive Trade Practices Act 1976

1976 c 34

Consolidates earlier restrictive trade practices legislation in England, Scotland and Wales. In particular, exempts from the provisions of the Act certain agreements of agricultural and/or forestry associations in connection with the marketing or preparation of produce produced by members of the associations on land occupied by them and used for agriculture or forestry (s33) [Certain limitations are imposed by the Agricultural and Forestry Associations Order 1974].

Finance Act 1976

1976 c 40

Makes an amendment to Schedule 9 to the Finance Act 1975 in its paragraph 6(2)(b) (relief for woodlands: expenses allowable under the Schedule to include those incurred in replanting within 3 years of a disposal) by extending this period to such longer time as the Commissioners of Inland Revenue may allow (s75).
Includes in Schedule 10 a general business relief from Capital Transfer Tax for which woodlands can qualify, as an alternative to the special relief granted under Schedule 9 of the Finance Act 1975. This relief reduces the value of the business by 30% before tax is charged.

Local Government (Miscellaneous Provisions) Act 1976

1976 c 57

Empowers local authorities in England and Wales to deal with dangerous trees; enables them to take such steps on land, whether by felling trees or otherwise, as are appropriate to make trees safe and makes provision for the recovery of expenses (s23 and 24).

Development of Rural Wales Act 1976

1976 c 75

Establishes a Development Board for Rural Wales. The Board is prohibited from carrying out forestry or afforestation on land held by it (s1(6)) except for the purpose of continuing existing forestry use of land which it acquires for development, prior to changing its use to another purpose (s4 (2)).

Rent (Agriculture) Act 1976

1976 c 80

Affords security of tenure for agricultural workers (including certain persons employed in forestry) in England and Wales housed by their employers. Temporary provisions are applied as respects certain forestry workers including the deferment of the operative date of the provisions of the Act (s1(5) and Part II of Schedule 3). Certain transitional provisions also apply to forestry workers (s40 and paragraph 7 of Schedule 9).

Forest Reproductive Material Regulations 1977

SI 1977/891

Control the marketing of forest reproductive material (seed, parts of plants and young plants) in Great Britain. Make provision for maintaining standards relating to genetic characters and external quality (these Regulations give effect to the appropriate EEC Directives).

Importation of Wood and Bark (Prohibition)(Great Britain) Order 1977

SI 1977/901

Prohibits the importation into Great Britain of certain kinds of wood and bark from specific countries unless certain plant health conditions have been fulfilled.

Finance Act 1977

Omits from Schedule 9 of the Finance Act 1975 the requirement that relief from Capital Transfer Tax under that Schedule should be conditional on the woodlands being managed in accordance with a plan approved by the Forestry Commissioners under a forestry dedication covenant or agreement.

Dutch Elm Disease (Restriction on Movement of Elms) Order 1977

SI 1977/1075

Revokes and re-enacts with modifications the provisions of the Dutch Elm Disease (Restriction on Movement of Elms) Order 1975 (as amended).
Restricts the movement in Great Britain of all elm by providing that, except as provided by Article 3(2) and (3), no elm shall be moved

- out of the areas listed in the Schedule or
- from any place within any area not listed in that Schedule to any other place by a highway, by rail or water,

unless the bark of the elm has been removed.
Empowers the Forestry Commissioners or an inspector, if satisfied that any elm is or has been moved in contravention of the Order, to serve a notice on the owner or person in charge of it requiring the removal of bark from the elm or the destruction of the elm. If the requirements of such a notice are not fulfilled, the necessary work may be carried out by an inspector and the expenses incurred recovered from the person in default.
In addition, gives certain powers of entry and examination to inspectors, prescribes penalties and provides for the furnishing of information and the production of records relating to diseased elms or elms in respect of which a contravention of the Order has taken place.

Dutch Elm Disease (Local Authorities) Order 1977

SI 1977/1074

Revokes and re-enacts with minor modifications the Dutch Elm Disease (Local Authorities) Order 1974 (as amended). Gives powers to local authorities listed in the Schedule to the order to take steps in connection with Dutch Elm disease. Gives officers appointed by these authorities powers to issue notices requiring the destruction of trees and the taking of steps to prevent the spread of the disease.

Index